International Mathematical Olympiad Volume II

Anthem Science, Technology and Medicine

International Mathematical Olympiad Volume II

1976–1990

ISTVAN REIMAN

Anthem Press

Anthem Press
An imprint of Wimbledon Publishing Company
75-76 Blackfriars Road, London SE1 8HA
or
PO Box 9779, London SW19 7ZG
www.anthempress.com

This edition published by Anthem Press 2005.
First published by Typotex Ltd in Hungarian as
Nemzetközi Matematikal Diákolimpiák by I Reiman.

Translated by János Pataki, András Stipsitz & Csaba Szabó

British Library Cataloguing in Publication Data
A catalogue record for this book is available from the British Library.

Library of Congress Cataloguing in Publication Data
A catalogue record for this book has been requested.

1 3 5 7 9 10 8 6 4 2

ISBN 1 84331 199 2 (Hbk)
ISBN 1 84331 200 X (Pbk)

Cover Illustration: Footprint Labs

Prelims typeset by Footprint Labs Ltd, London
www.footprintlabs.com

Printed in India

Preface

This three volume set contains the problems from the first forty-five IMO-s, from 1959 to 2004.

The chronicle of the IMO (International Mathematics Olympiad) starts with the initiative of the Romanian Mathematics and Physics Society: in July 1959 on the occasion of a celebration the Society invited high school students from the neighbouring countries to an international mathematical competition. The event proved to be such a success that the participants all agreed to go on with the enterprise. Ever since, this competition has taken place annually (except for 1980) and it has gradually transformed from the local contest of but a few countries into the most important and comprehensive international mathematical event for the young. Only seven nations were invited for the first IMO, while the number of participating countries was well beyond eighty for the last event; wherever mathematical education has reached a moderate level, sooner or later the country has turned up at the IMO.

The movement has had a significant impact on the mathematical education of several participating countries and also on the development of the gifted. The aim of a more proficient preparation for the IMO itself has launched the organization of national mathematical competitions in many countries involved. As the crucial component of successful participation, the preparation for the contest has enriched the publishing activity in several countries. Math-clubs have been formed on a large scale and periodicals have started. Even though the competition certainly brings up some pedagogical problems, if the educators regard the competitions not as ultimate aims, but as ways to introduce and endear pupils to mathematics, then their pedagogical benefit is undeniable.

The administration of the competitions has not changed that much; the larger scale has obviously necessitated certain modifications but the actual contest is more or less as it used to be. The participating countries are invited to delegate a group of up to six students who are attending high school at the year of the contest or had just finished their secondary school studies. Three problems are posed each day over two consecutive days and the students have to produce written solutions in their native tongue. There are two delegation leaders accompanying each team; one of their tasks is to provide an oral translation of their students' work into one of the official languages—by now this has been almost exclusively English—for a committee of mathematicians from the host country. Together with this group of coordinators they eventually settle the score the solutions are worth; the highest mark is seven points for each problem. The contestants are then ranked according to their total scores; the awarding of the prizes has been administered according to the following principle: half of the participants are given a prize: namely the proportion of the gold, silver and bronze medals is $1:2:3$ respectively.

The occasional professional problems are handled by the international jury formed by the leaders of the participating delegations; their most important and difficult task is to select the six problems for the actual contest, to formulate their official text and to prepare rough marking schemes for each of them. The organizers ask for proposals from the participating countries well in advance; in due course they produce a list of approximately twenty to twenty five problems from those suggested and the jury selects the final six from this supply.

There are almost two hundred problems in this three volume set and they provide a full image of the challenge the students had to cope with during these forty years. One cannot claim that every single one of them is a pearl of mathematics but their overwhelming majority is interesting and rewarding; together they more or less cover the usual syllabus-chapters of elementary mathematics. When selecting, the jury usually tries to choose from the intersection of the respective curricula of the participating countries; considering that there are more than eighty of them this is not an easy job, if not impossible. The reader might notice that there are no problems at all from the theory of probability, for example, and complex numbers hardly show up.

From the retrospect of more than forty years one can certainly conclude that the IMO movement has had a significant role in the history of the second half of twentieth century mathematics. There are quite a few highly ranked mathematicians who started their career at an IMO; even at this point, however, we have to emphasize, that an eventual fiasco at the IMO or any other mathematical contest whatsoever usually has no implications at all about the mathematical potential of a well prepared student.

A careful reader will certainly realize that quite a few problems in this book are in fact simplifications or particular cases of more profound mathematical results; apart from the intellectual satisfaction of actually solving these problems, the discovery of this mathematical background and the knowledge gained from it can be the ultimate benefits of a high level study of this book.

At the end of the book we included a Glossary of Theorems (and their proofs) we used in the book and we refer to them by their numbers enclosed in brackets, e. g. [6].

István Reiman

International Mathematical Olympiad

Problems

1976

1976/1. A plane convex quadrilateral has area 32 cm^2, and the sum of two opposite sides and a diagonal is 16 cm. Determine all possible lengths for the other diagonal.

1976/2. Let $P_1(x) = x^2 - 2$; $P_j(x) = P_1(P_{j-1}(x))$; $j = 2, 3, \ldots$. Prove that for every n, $P_n(x) = x$ has n distinct real roots.

1976/3. A rectangular box can be completely filled with unit cubes. If one places as many cubes as possible, each with volume 2, in the box, with their edges parallel to the edges of the box, one can fill exactly 40% of the box. Determine the possible dimensions of the box. ($\sqrt[3]{2} = 1.2599\ldots$).

1976/4. Determine the largest number which is the product of positive integers with sum 1976.

1976/5. Consider the following system of p equations with $q = 2p$ unknowns:

$$a_{11}x_1 + a_{12}x_2 + \ldots + a_{1q}x_q = 0,$$
$$a_{21}x_1 + a_{22}x_2 + \ldots + a_{2q}x_q = 0,$$
$$\ldots\ldots\ldots\ldots\ldots\ldots\ldots\ldots\ldots$$
$$a_{p1}x_1 + a_{p2}x_2 + \ldots + a_{pq}x_q = 0,$$

where $a_{ij} \in \{0, 1, -1\}$ ($i = 1, 2, \ldots, p$, $j = 1, 2, \ldots, q$).

Prove that the system of equations has a solution $x_1, x_2, \ldots, x_q$ satisfying the following properties:

a) $x_1, x_2, \ldots, x_q$ are integers;

b) not all x_j are 0 ($j = 1, 2, \ldots, q$);

c) $|x_j| \leq q$ ($j = 1, 2, \ldots, q$).

1976/6. The sequence $u_0, u_1, \ldots$ is defined as follows: $u_0 = 2$, $u_1 = \dfrac{5}{2}$, $u_{n+1} = u_n(u_{n-1}^2 - 2) - u_1$ ($n = 1, 2, \ldots$). Prove that

$$u_n = 2^{\frac{2^n - (-1)^n}{3}} \qquad (n = 1, 2, \ldots).$$

1977

1977/1. Construct equilateral triangles ABK, BCL, CDM, DAN on the inside of the square $ABCD$. Show that the midpoints of KL, LM, MN, NK and the midpoints of AK, BK, BL, CL, CM, DM, DN, AN form a regular dodecahedron.

1977/2. In a finite sequence of real numbers the sum of any seven consecutive terms is negative, and the sum of any eleven consecutive terms is positive.

Determine the maximum number of terms in the sequence.

1977/3. Given an integer $n > 2$, let V_n be the set of integers $1 + kn$ for k a positive integer. A number $m \in V_n$ is called indecomposable if there is no number $p, q \in V_n$ such that $pq = m$.

Prove that there is a number r in V_n which can be expressed as the product of indecomposable members of V_n in more than one way (decompositions which differ solely in the order of factors are not regarded as different).

1977/4. Define

(1) $$f(x) = 1 - a\cos x - b\sin x - A\cos 2x - B\sin 2x$$

where a, b, A and B are real constants. Suppose that $f(x) \geq 0$ for all real x. Prove that

(2) $$a^2 + b^2 \leq 2, \qquad A^2 + B^2 \leq 1.$$

1977/5. Let a and b be positive integers. When $(a^2 + b^2)$ is divided by $(a + b)$, the quotient is q, the remainder r.

Find all pairs (a, b), such that

(1) $$q^2 + r = 1977.$$

1977/6. The function f is defined on the set of positive integers and its values are positive integers. Let us assume that

(1) $$f(n+1) > f(f(n))$$

for every positive n.

Prove that for every positive n

$$f(n) = n.$$

1978

1978/1. For m and n positive integers, $n > m > 1$, the last three decimal digits of 1978^m is the same as the last three decimal digits of 1978^n. Find m and n such that $m + n$ has the least possible value.

1978/2. P is a point inside a sphere. Three mutually perpendicular rays from P intersect the sphere at points U, V and W. Let Q denote the vertex diagonally opposite P in the parallelepiped determined by PU, PV, PW. Find the locus of Q for all possible sets of such rays from P.

1978/3. The set of all positive integers is the union of two disjoint subsets:
$$F = \{f(1), f(2), \ldots, f(n), \ldots\} \quad \text{and}$$
$$G = \{g(1), g(2), \ldots, g(n), \ldots\}$$
where
$$f(1) < f(2) < \ldots < f(n) < \ldots \quad \text{and} \quad g(1) < g(2) < \ldots < g(n) < \ldots,$$
(1) $$g(n) = f(f(n)) + 1 \quad \text{for every } n \geq 1.$$
Determine $f(240)$.

1978/4. In the triangle ABC we have $AB = AC$. A circle is tangent internally to the circumcircle of the triangle and also to sides AB, AC at P, Q respectively. Prove that the midpoint of PQ is the centre of the incircle of the triangle.

1978/5. Let $\{a_k\}$ be a sequence of distinct positive integers ($k = 1, 2, \ldots, n, \ldots$). Prove that for every positive integer n
(1) $$\sum_{k=1}^{n} \frac{a_k}{k^2} \geq \sum_{k=1}^{n} \frac{1}{k}.$$

1978/6. An international society has its members from six different countries. The list of members has 1978 names, numbered 1, 2, ..., 1978. Prove that there is at least one member whose number is the sum of the numbers of two members from his own country, or twice the number of a member from his own country.

1979

1979/1. Let p and q be positive integers such that
$$\frac{p}{q} = 1 - \frac{1}{2} + \frac{1}{3} - \frac{1}{4} + \ldots - \frac{1}{1318} + \frac{1}{1319}.$$
Prove that 1979 divides p.

1979/2. A prism with pentagons $A_1A_2A_3A_4A_5$, and $B_1B_2B_3B_4B_5$ as top and bottom faces is given. Each side of the two pentagons and each of the 25 segments A_iB_j, ($i, j = 1, 2, \ldots, 5$) is coloured red or green. Every triangle whose vertices are vertices of the prism and whose sides have all been coloured has two sides of a different colour. Prove that all 10 edges of the top and bottom faces have the same colour.

1979/3. Let k_1 and k_2 be two circles on the plane and let A denote one of their points of intersection. Starting simultaneously from A, two points, P_1 and P_2 move with constant speed, each traveling along its own circle in the same sense. The two points return to A simultaneously after one revolution. Prove that there is a fixed point P in the plane such that, at any time, the distance from P to the moving points are equal.

1979/4. Given a plane π, a point P in the plane and a point Q not in the plane. Find all points R of the plane π such that the ratio

(1)
$$\frac{QP + PR}{QR}$$

is maximal.

1979/5. Find all real numbers b for which there exist nonnegative real numbers x_1, x_2, x_3, x_4, x_5 satisfying

(1)
$$\sum_{k=1}^{5} k x_k = b, \quad \sum_{k=1}^{5} k^3 x_k = b^2, \quad \sum_{k=1}^{5} k^5 x_k = b^3.$$

1979/6. Let A and E be opposite vertices of an octagon. A frog starts at vertex A. From any vertex except E it jumps to one of the two adjacent vertices. When it reaches E it stops. Let a_n be the number of distinct paths of exactly n jumps ending at E. Prove that

$$a_{2n-1} = 0, \qquad a_{2n} = \frac{1}{\sqrt{2}}\left(x^{n-1} - y^{n-1}\right) \qquad (n = 1,\ 2,\ 3,\ \ldots)$$

where $x = 2 + \sqrt{2}$ and $y = 2 - \sqrt{2}$.

1980

In 1980 there was no International Mathematical Olympics. The Organisation of the Teachers of Mathematics, Physics and Chemistry of Finland (MAOL) invited the teams of England, Hungary and Sweden to compare their abilities. Here we present the problems of that competition.

1980/1. Let α, β, and γ denote the angles of the triangle ABC. The perpendicular bisector of AB intersects BC at the point X, the perpendicular bisector of AC intersects it at Y. Prove that $\tan\beta \cdot \tan\gamma = 3$ implies $BC = XY$.

Show that this condition is not necessary for $BC = XY$, and give a sufficient and necessary condition.

1980/2. Define the numbers $a_0, a_1, \ldots, a_n$ in the following way:

$$a_0 = \frac{1}{2}, \quad a_{k+1} = a_k + \frac{a_k^2}{n} \qquad (n > 1, \ k = 0, 1, \ldots, n-1).$$

Prove that

(1)
$$1 - \frac{1}{n} < a_n < 1.$$

1980/3. Prove that the equation

(1)
$$x^n + 1 = y^{n+1},$$

where n is a positive integer not smaller then 2, has no positive integer solution in x and y for which x and $n+1$ are relatively prime.

1980/4. Determine all positive integers n such that the following statement holds: In the inscribed convex polygon $A_1 A_2 \ldots A_{2n}$ if the pairs of opposite sides

$$(A_1 A_2, A_{n+1} A_{n+2}), \ (A_2 A_3, A_{n+2} A_{n+3}), \ \ldots, \ (A_{n-1} A_n, A_{2n-1} A_{2n})$$

are parallel, then the sides

$$A_n A_{n+1}, A_{2n} A_1$$

are parallel as well.

1980/5. In a rectangular coordinate system we call a line parallel to the x axis triangular if it intersects the curve with equation

$$y = x^4 + px^3 + qx^2 + rx + s$$

in the points A, B, C and D (from left to right) such that the segments AB, AC and AD are the sides of a triangle.

Prove that the lines parallel to the x axis intersecting the curve in four distinct points are all triangular or none of them is triangular.

1980/6. Find the digits left and right of the decimal point in the decimal form of the number

$$\left(\sqrt{2} + \sqrt{3}\right)^{1980}.$$

1981

1981/1. Let P be a point inside the triangle ABC and D, E, F are the feet of the perpendiculars from P to the lines BC, CA, AB, respectively. Find all P which minimise:

(1)
$$\frac{BC}{PD} + \frac{CA}{PE} + \frac{AB}{PF}.$$

1981/2. Consider all subsets of size r of the set $H_n = \{1, 2, \ldots, n\}$, where $1 \leq r \leq n$. Each subset has a minimal element, let $F(n, r)$ denote the arithmetic mean of these elements. Prove that

$$F(n, r) = \frac{n+1}{r+1}.$$

1981/3. Determine the maximum value of $m^2 + n^2$ where m and n are integers satisfying $m, n \in \{1, 2, \ldots, 1981\}$ and

(1) $$(n^2 - nm - m^2)^2 = 1.$$

1981/4. a) For which $n > 2$ is there a set of n consecutive positive integers such that the largest number in the set is a divisor of the least common multiple of the remaining $n - 1$ numbers?

b) For which $n > 2$ is there exactly one set having this property?

1981/5. Three circles of equal radii have a common point O and lie inside a given triangle. Each circle touches a pair of sides of the triangle. Prove that the incentre and the circumcentre of the triangle are collinear with the point O.

1981/6. The function $f(x, y)$ satisfies

(1) $$f(0, y) = y + 1,$$

(2) $$f(x + 1, 0) = f(x, 1),$$

(3) $$f(x + 1, y + 1) = f(x, f(x + 1, y))$$

for every integer x and y. Find $f(4, 1981)$.

1982

1982/1. The function $f(n)$ is defined on the positive integers and takes on non-negative integer values. For all n, m

(1) $$f(m+n) - f(m) - f(n) = 0 \quad \text{or} \quad 1,$$

(2) $$f(2) = 0, \qquad f(3) > 0,$$

(3) $$f(9999) = 3333.$$

Determine $f(1982)$.

1982/2. A non-isosceles triangle $A_1 A_2 A_3$ has sides a_i opposite to A_i. M_i is the midpoint of side a_i and T_i is the point where the incircle touches side a_i.

Denote by S_i the reflection of T_i in the interior bisector of angle A_i. Prove that the lines $M_1 S_1$, $M_2 S_2$ and $M_3 S_3$ are concurrent.

1982/3. Consider the infinite non increasing sequence $\{x_i\}$ of positive reals such that $x_0 = 1$.

a) Prove that for every such sequence there is an $n \geq 1$, such that

(1)
$$S_n = \frac{x_0^2}{x_1} + \frac{x_1^2}{x_2} + \ldots + \frac{x_{n-1}^2}{x_n} \geq 3{,}999.$$

b) Find such a sequence for which

$$S_n = \frac{x_0^2}{x_1} + \frac{x_1^2}{x_2} \ldots + \frac{x_{n-1}^2}{x_n} < 4.$$

for all n.

1982/4. Prove that if n is a positive integer such that the equation
$$x^3 - 3xy^2 + y^3 = n$$
has a solution in integers x, y, then it has at least three such solutions. Show that the equation has no solutions in integers for $n = 2891$.

1982/5. The diagonals AC and CE of the regular hexagon $ABCDEF$ are divided by inner points M and N respectively, so that:

(1)
$$\frac{AM}{AC} = \frac{CN}{CE} = r.$$

Determine r, if B, M and N are collinear.

1982/6. Let S be a square with sides length 100. Let L be a path within S which does not meet itself and which is composed of line segments A_0A_1, A_1A_2, $\ldots$, $A_{n-1}A_n$, where $A_0 \neq A_n$.

Suppose that for every point P on the boundary of S there is a point of L of distance from P no greater than $1/2$. Prove that there are two points X and Y of L such that the distance between X and Y is not greater than 1 and the length of the part of L which lies between X and Y is not smaller than 198.

1983

1983/1. Find all functions f defined on the set of positive reals which take positive real values and satisfy:

(1)
$$f(xf(y)) = yf(x)$$

for every positive x and y. Show that

(2)
$$f(x) \to 0, \quad \text{if} \quad x \to \infty.$$

1983/2. Let A be one of the two distinct points of intersection of two unequal coplanar circles C_1 and C_2 with centres O_1 and O_2, respectively. One of the common tangents to the circles touches C_1 at P_1, C_2 at P_2 while the other touches C_1 at Q_1 and C_2 at Q_2. Let M_1 be the midpoint of P_1Q_1 and M_2 be the midpoint of P_2Q_2. Prove that $\angle O_1AO_2 = \angle M_1AM_2$.

1983/3. Let a, b, c denote pairwise coprime positive integers. Prove that
$$2abc - ab - bc - ca$$
is the largest integer which cannot be expressed as

(1) $$xbc + yca + zab,$$

where x, y, z are non negative integers.

1983/4. Let ABC be an equilateral triangle and E the set of all points contained in the three segments AB, BC and CA (including A, B and C). Determine whether, for every partition of E into two disjoint subsets, at least one of the two subsets contains the vertices of a right-angled triangle.

1983/5. Is it possible to choose 1983 distinct positive integers, all less than or equal to 10^5, no three of which are consecutive terms of an arithmetic progression?

1983/6. Let a, b, c be the length of the sides of a triangle. Prove that

(1) $$a^2b(a - b) + b^2c(b - c) + c^2a(c - a) \geq 0.$$

1984

1984/1. Prove that

(1) $$0 \leq xy + yz + zx - 2xyz \leq \frac{7}{27},$$

where x, y and z are nonnegative real numbers for which

(2) $$x + y + z = 1.$$

1984/2. Find one pair of positive integers, a, b, such that:
(1) $ab(a + b)$ is not divisible by 7;
(2) $(a + b)^7 - a^7 - b^7$ is divisible by 7^7.

1984/3. In the plane two different points O and A are given. For each point $X \neq O$ on the plane denote by $\omega(X)$ the measure of the angle between OA and OX in radians counterclockwise from OA $(0 \leq \omega(X) < 2\pi)$. Let $C(X)$ be the circle with centre O and radius $OX + \dfrac{\omega(X)}{OX}$. Each point of the plane is coloured by one of a finite number of colours. Prove that there exists a point Y for which $\omega(Y) > 0$ such that its colour appears on the circumference of the circle $C(Y)$.

1984/4. Let $ABCD$ be a convex quadrilateral with the line CD tangent to the circle on diameter AB. Prove that the line AB is tangent to the circle on diameter CD if and only if BC and AD are parallel.

1984/5. Let d be the sum of the lengths of all the diagonals of a plane convex polygon with $n \geq 3$ vertices. Let p be its perimeter. Prove that:

(1)
$$n - 3 < \frac{2d}{p} < \left[\frac{n}{2}\right]\left[\frac{n+1}{2}\right] - 2.$$

1984/6. Let a, b, c, d be odd numbers, such that:

(1) $$0 < a < b < c < d,$$

(2) $$ad = bc,$$

(3) $$a + d = 2^k, \qquad b + c = 2^m$$

holds, where k and m are integers. Prove that $a = 1$.

1985

1985/1. A circle has centre on the side AB of the cyclic quadrilateral $ABCD$. The other three sides are tangents to the circle. Prove that

$$AD + BC = AB.$$

1985/2. Let n and k be relatively prime positive integers with $k < n$. Each number in the set $M = \{1, 2, 3, \ldots, n-1\}$ is coloured either blue or white, such that:

(a) for each i in M, both i and $n - i$ have the same colour;

(b) for each i in M not equal to k, both i and $|i - k|$ have the same colour.

Prove that all numbers in M must have the same colour.

1985/3. For any polynomial $P(x) = a_0 + a_1 x + \ldots + a_k x^k$ with integer coefficients let $\omega(P)$ denote the number of odd coefficients, and let $Q_i(x) = (1+x)^i$, where $i = 0, 1, 2, \ldots$.

Prove that if $i_1, i_2, \ldots, i_n$ are integers such that $0 \leq i_1 < i_2 < \ldots < i_n$, then

(1)
$$\omega\left(Q_{i_1} + Q_{i_2} + \ldots + Q_{i_n}\right) \geq \omega\left(Q_{i_1}\right).$$

1985/4. Given a set M of 1985 distinct positive integers, none of which has a prime divisor greater than 26. Prove that M contains a subset of 4 elements whose product is the 4th power of an integer.

1985/5. The circle k_1 with centre O passes through the vertices A and C of the triangle ABC and intersects the segments AB and BC again at distinct points K and N, respectively. The circumcircles k of ABC and k_2 of KBN intersect at exactly two distinct points B and M. Prove that $\angle O_1 M B = 90°$.

1985/6. For every real number x_1 construct the sequence x_1, x_2, ..., x_n, where

(1)
$$x_{n+1} = x_n \left(x_n + \frac{1}{n} \right)$$

for every $n \geq 1$. Prove that there exists exactly one value of x_1 which gives
$$0 < x_n < x_{n+1} < 1$$
for every positive n.

1986

1986/1. Let d be any positive integer not equal to 2, 5 or 13. Show that one can find distinct a, b in the set $\{2, 5, 13, d\}$ such that $ab - 1$ is not a perfect square.

1986/2. Given a point P in the plane of the $A_1 A_2 A_3$ triangle. Define $A_s = A_{s-3}$ for $s \geq 4$.

Construct a series of points P_0, P_1, P_2, ... such that P_{k+1} is the image of P_k under a rotation with centre A_{k+1} through an angle $-120°$ ($k = 0, 1, 2, ...$).

Prove that if $P_{1986} = P_0$, then the triangle $A_1 A_2 A_3$ is equilateral.

1986/3. To each vertex of a regular pentagon an integer is assigned, so that the sum of all five numbers is positive. If three consecutive vertices are assigned the numbers x, y, z respectively, and $y < 0$, then the following operation is allowed:

x, y, z are replaced by $x+y$, $-y$, $z+y$ respectively. Such an operation is performed repeatedly as long as at least one of the five numbers is negative. Determine whether this procedure necessarily comes to an end after a finite number of steps.

1986/4. Let A, B be adjacent vertices of a regular n-gon ($n \geq 5$) with centre O. A triangle XYZ, which is congruent to and initially coincides with OAB, moves in the plane in such a way that Y and Z each trace out the whole boundary of the polygon, with X remaining inside the polygon. Find the locus of X.

1986/5. Find all functions f defined on the non-negative reals and taking non-negative real values such that:

(a) $f(x \cdot f(y)) \cdot f(y) = f(x+y)$ for every non negative x and y;

(b) $f(2) = 0$;

(c) $f(x) \neq 0$, if $0 \leq x < 2$.

1986/6. Given a finite set of points in the plane, each with integer coordinates. Is it always possible to colour the points red or white so that for any straight line L parallel to one of the coordinate axes the difference (in absolute value) between the numbers of white and red points on L is not greater than 1?

1987

1987/1. Let $p_n(k)$ be the number of permutations of the set $S = \{1, 2, \ldots \ldots, n\}$ $(n \geq 1)$ which have exactly k fixed points. Prove that the sum from $k = 0$ to n of $kp_n(k)$ is $n!$.

1987/2. In an acute-angled triangle ABC the interior bisector of angle A meets BC at L and meets the circumcircle of ABC again at N. From L perpendiculars are drawn to AB and AC, with feet K and M respectively. Prove that the quadrilateral $AKNM$ and the triangle ABC have equal areas.

1987/3. Let $x_1, x_2, \ldots, x_n$ be real numbers satisfying

(1)
$$x_1^2 + x_2^2 + \ldots + x_n^2 = 1.$$

Prove that for every integer $k \geq 2$ there are integers a_i $(i = 1, 2, \ldots, n)$, not all zero, such that $|a_i| \leq k - 1$ for all i, and

(2)
$$|a_1 x_1 + a_2 x_2 + \ldots + a_n x_n| \leq \frac{(k-1)\sqrt{n}}{k^n - 1}.$$

1987/4. Prove that there is no function f from the set of non-negative integers into itself such that

(1)
$$f(f(n)) = n + 1987$$

for all n.

1987/5. Let n be an integer greater or equal to 3. Prove that there is a set of n points in the plane such that the distance between any two points is irrational and each set of 3 points determines a non-degenerate triangle with rational area.

1987/6. Let n be an integer greater or equal to 2. Prove that if $k^2 + k + n$ is prime for all integers such that $0 \leq k \leq \sqrt{\dfrac{n}{3}}$, then $k^2 + k + n$ is prime for all integers k such that $0 \leq k \leq n - 2$.

1988

1988/1. Consider two coplanar circles of radii R and r $(R > r)$ with the same centre. Let P be a fixed point on the smaller circle and B a variable point on the larger circle. The line BP meets the larger circle again at C. The perpendicular l to BP at P meets the smaller circle again at A (if l is tangent to the circle at P then $A = P$).

(i) Find the set of values of $BC^2 + CA^2 + AB^2$.

(ii) Find the locus of the midpoint of AB.

1988/2. Let n be a positive integer and let A_1, A_2, ..., A_{2n+1} be subsets of a set B. Suppose that

a) each A_i has exactly $2n$ elements,

b) each $A_i \cap A_j$ $(1 \leq i < j \leq 2n+1)$ contains exactly one element, and

c) every element of B belongs to at least two of the A_i.

For which values of n can one assign to every element of B one of the numbers 0 and 1 in such a way that each A_i has 0 assigned to exactly n of its elements?

1988/3. A function f is defined on the positive integers by

(1) $$f(1) = 1, \qquad f(3) = 3,$$
(2) $$f(2n) = n,$$
(3) $$f(4n+1) = 2f(2n+1) - f(n),$$
(4) $$f(4n+3) = 3f(2n+1) - 2f(n)$$

for all positive integers n.

Determine the number of positive integers n, less than or equal to 1988, for which $f(n) = n$.

1988/4. Show that the set of real numbers x which satisfy the inequality

(1) $$\sum_{k=1}^{70} \frac{k}{x-k} \geq \frac{5}{4}$$

is a union of disjoint intervals, the sum of whose lengths is 1988.

1988/5. ABC is a triangle right-angled at A, and D is the foot of the altitude from A. The straight line joining the incentres of the triangles ABD, ACD intersects the sides AB, AC at the points K, L respectively. S and T denote the areas of the triangles ABC and AKL respectively. Show that $S \geq 2T$.

1988/6. Let a and b be positive integers such that $ab+1$ divides a^2+b^2. Show that

(1) $$\frac{a^2+b^2}{ab+1}$$

is the square of an integer.

1989

1989/1. Prove that the set $\{1, 2, \ldots, 1989\}$ can be expressed as the disjoint union of subsets A_1, A_2, ..., A_{117} in such a way that each A_i contains 17 elements and the sum of the elements in each A_i is the same.

1989/2. In an acute-angled triangle ABC, the internal bisector of angle A meets the circumcircle again at A_1. Points B_1 and C_1 are defined similarly. Let A_0 be the point of intersection of the line AA_1 with the external bisectors of angles B and C. Points B_0 and C_0 are defined similarly. Prove that the area of the triangle $A_0B_0C_0$ is twice the area of the hexagon $AC_1BA_1CB_1$ and at least four times the area of the triangle ABC.

1989/3. Let n and k be positive integers and let S be a set of n points in the plane such that no three points of S are collinear, and for any point P of S there are at least k points of S equidistant from P. Prove that

$$k < \frac{1}{2} + \sqrt{2n}.$$

1989/4. Let $ABCD$ a convex quadrilateral such that the sides AB, BC, AD satisfy $AB = AD + BC$. There exists a point P inside the quadrilateral at a distance h from the line CD such that $AP = h + AD$ and $BP = h + BC$. Show that

(1) $$\frac{1}{\sqrt{h}} \geq \frac{1}{\sqrt{AD}} + \frac{1}{\sqrt{BC}}.$$

1989/5. Prove that for each positive integer n there exist n consecutive positive integers none of which is a prime or a prime power.

1989/6. A permutation x_1, x_2, $\ldots$, x_{2n-1}, x_{2n} of the set 1, 2, $\ldots$, $2n - 1$, $2n$ where n is a positive integer is said to have property P if $|x_i - x_{i+1}| = n$ for at least one i in $\{1, 2, \ldots, 2n - 1\}$. Show that for each n there are more permutations with property P than without.

1990

1990/1. Chords AB and CD of a circle intersect at a point E inside the circle. Let M be an interior point of the segment EB. The tangents to the circle through D, E and M intersects the lines BC and AC at F and G respectively. Find $\dfrac{EG}{EF}$ in terms of $t = \dfrac{AM}{AB}$.

1990/2. Take $n \geq 3$ and consider a set E of $2n - 1$ distinct points on a circle. Suppose that exactly k of these points are to be coloured black. Such a colouring is "good" if there is at least one pair of black points such that the interior of one of the arcs between them contains exactly n points from E. Find the smallest value of k so that every such colouring of k points of E is good.

1990/3. Determine all integers greater than 1 such that $\dfrac{2^n + 1}{n^2}$ is an integer.

1990/4. Construct a function from the set of positive rational numbers into itself such that

(1)
$$f(x \cdot f(y)) = \frac{f(x)}{y}$$

for all x, y.

1990/5. Given an initial integer $n_0 > 1$, two players A and B choose integers n_1, n_2, n_3, ... alternately according to the following rules: knowing n_{2k}, A chooses any integer n_{2k+1} such that

$$n_{2k} \le n_{2k+1} \le n_{2k}^2.$$

Knowing n_{2k+1}, B chooses any integer n_{2k+2} such that

$$\frac{n_{2k+1}}{n_{2k+2}} = p^r$$

for some prime p and integer $1 \le r$. Player A wins the game by choosing the number 1990; player B wins by choosing number 1. For which n_0 does

a) A have a winning strategy?

b) B have a winning strategy?

c) Neither player have a winning strategy?

1990/6. Prove that there exists a convex 1990-gon such that all its angles are equal and the lengths of the sides are the numbers 1^2, 2^2, ..., 1989^2, 1990^2 in some order.

Solutions

1976.

1976/1. *A plane convex quadrilateral has area* 32 cm^2, *and the sum of two opposite sides and a diagonal is* 16 cm. *Determine all possible lengths for the other diagonal.*

Solution. Let $ABCD$ be the convex quadrilateral and $AB = a$, $CD = c$, $AC = f$, $BD = e$ (see *Figure 1976/1.1*). Let F_{ABC} denote the area of the ABC triangle. By the assumptions $a + e + c = 16$.

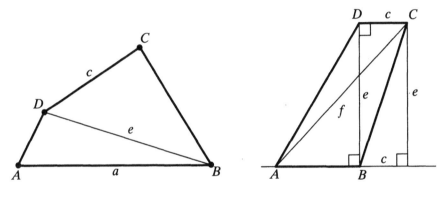

Figure 76/1.1   Figure 76/1.2

Twice the area of a triangle is smaller or equal than the product of its sides, and is equal to it if and only if the two sides are orthogonal. Thus

(1) $$2 \cdot 32 = 2F_{ABD} + 2F_{BDC} \leq ae + ec = (a + c)e = (16 - e)e,$$

hence

(2) $$0 \geq e^2 - 16e + 64 = (e - 8)^2,$$

and this holds only if $e = 8$ and $a + c = 8$. In case of equality in (1) and (2) AB and CD are orthogonal to the BD diagonal (see *Figure 1976/1.2*). Thus the only possible length of the other diagonal is:

$$f = \sqrt{(a + c)^2 + e^2} = \sqrt{8^2 + 8^2} = 8\sqrt{2}.$$

1976/2. *Let* $P_1(x) = x^2 - 2$; $P_j(x) = P_1(P_{j-1}(x))$; $j = 2, 3, \ldots$. *Prove that for every* n, $P_n(x) = x$ *has* n *distinct real roots.*

First solution. As the degree of the polynomial P_j is twice the degree of P_{j-1}, $(j = 2, 3, \ldots)$, the polynomial $P_n(x) - x$ has degree 2^n and has at most

2^n real roots. Let us examine the roots in the interval $[-2, 2]$; its points can be written in the form $x = 2\cos t$. As $\cos 2t = 2\cos^2 t - 1$,

(1) $$P_1(x) = P_1(2\cos t) = 4\cos^2 t - 2 = 2\cos 2t.$$

Using induction we get

(2) $$P_n(x) = P_n(2\cos t) = 2\cos 2^n t.$$

Indeed, for $n = 1$ our statement is the same as (1). Assume that $P_{n-1}(x) = 2\cos 2^{n-1}t$, then

$$P_n(x) = (P_{n-1}(x))^2 - 2 = 4\cos^2 2^{n-1}t - 2 = 2(2\cos^2 2^{k-1}t - 1) = 2\cos 2^n t,$$

and we proved (2).

We can write the equation in the form

$$\cos 2^n t = \cos t$$

and hence

$$2^n t = t + 2k\pi \quad \text{or} \quad 2^n t = -t + 2k\pi, \quad k \text{ arbitrary integer}$$

and so

(3) $$t_1 = \frac{2k_1\pi}{2^n - 1}, \quad \text{and} \quad F_2 = \frac{2k_2\pi}{2^n + 1}.$$

If in case t_1 we have $k_1 = 0, 1, \ldots, 2^{n-1} - 1$ and in case t_2, $k_2 = 1, 2, \ldots, 2^{n-1}$, then we get the 2^n roots of the form $x = 2\cos t$. The values corresponding to t_1 are all distinct, and their cosine values are distinct, too, because the possible values of t_1 are in the interval $[0, \pi]$ where the $\cos x$ function is strictly increasing. The same holds for the t_2 values. It only remains to show that no t_1 and t_2 values agree. This would mean that there are integers k_1 and k_2 such that

$$\frac{2k_1\pi}{2^n - 1} = \frac{2k_2\pi}{2^n + 1},$$

$$k_1(2^n + 1) = k_2(2^n - 1)$$

held, as $2^n + 1$ and $2^n - 1$ are coprime and $2^n + 1$ cannot divide k_2. Hence the 2^n roots given in (3) are all distinct and so these are all the roots of the equation.

Second solution. The problem does not ask to determine the roots, only to prove their existence. Observe that if $P_{n-1}(x)$ attains the values $-2, 0, 2$ then at the same places the values of

$$P_n(x) = (P_{n-1}(x))^2 - 2$$

are $2, -2, 2$, respectively.

Using induction we prove the following property of $P_n(x)$: from -2 to $+2$, the polynomial $P_n(x)$ attains the values $+2$ and -2 alternating, it attains $+2$ at $2^{n-1} + 1$ many places and -2 at 2^{n-1} many places.

For $n = 1$ our statement is true as $P_1(-2) = 2$, $P_1(0) = -2$, $P_1(2) = 2$. Now, assume that it holds for $P_{n-1}(x)$, that is from -2 to $+2$ it attains $+2$ at $2^{n-2} + 1$ many places and -2 at 2^{n-2} many places, alternating. These points split the interval between the first and last examined points into 2^{n-1} parts. $P_{n-1}(x)$ equals to -2 at one end, and to $+2$ at the other end of these intervals. Because of the continuity of $P_{n-1}(x)$, it has a root in the interval. Now, joining these roots to the examined points, the values attained at these places are

$$2, \ 0, \ -2, \ 0, \ 2, \ -2, \ \ldots, \ -2, \ 0, \ 2,$$

respectively. Between every $(2, -2)$ and $(-2, 2)$ pair there is a 0, hence there are 2^{n-1} 0-s in the series. By our introductory remark, the values of $P_n(x)$ at these 0 places are

$$2, \ -2, \ 2, \ -2, \ 2, \ -2, \ \ldots, \ 2, \ -2, \ 2,$$

that is, there are $2^{n-1}+1$ 2-s and 2^{n-1} -2-s, and we proved our statement.

The statement of the problem easily follows, as these 2^{n-1} points split the interval $[-2, 2]$ into 2^n subintervals that can only intersect at the endpoints. At one end of these intervals $P_n(x) - x$ is positive and at the other and it is negative (except for $x = 2$). Hence, inside these intervals $P_n(x) - x$ has a root that is a solution of the equation $P_n(x) = x$. In the last interval, at one end the value is negative, at the other end non-negative, so there is a root, too. Thus $P_n(x) - x$ has at least 2^n distinct real roots, and it cannot have more roots since 2^n is the degree of $P_n(x) - x$.

Remarks. 1. The last step of the second solution says the following: as $P_n(x)$ attains the value 2 at $2^{n-1}+1$ places and the value -2 at 2^{n-1} many places in $[-2, 2]$, the curve $y = P_n(x)$ and the line $y = x$ intersect in at least 2^n points, at the roots of the equation $P_n(x) = x$ (see *Figure 1976/2.1*).

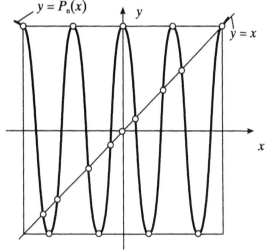

Figure 1976/2.1

2. Rewriting $P_n(x) = x$ to $(P_{n-1}(x))^2 - 2 = x$ we get

$$P_{n-1}(x) = \pm\sqrt{2+x};$$

continuing the procedure we get:

$$P_{n-2}(x) = \pm\sqrt{2 \pm \sqrt{2+x}}$$

$$\vdots$$

(2) $$P_1(x) = \pm\sqrt{2 \pm \sqrt{2 \pm \ldots \pm \sqrt{2+x}}} = x^2 - 2$$

$$x = \pm\sqrt{2 \pm \sqrt{2 \pm \ldots \pm \sqrt{2+x}}}.$$

This last form is, in fact, 2^n equations because that is the number of the possible combinations of the signs in the square roots. We get a solution to our equation if we fix a combination of the signs in (2). This means that the values $x = 2\cos t$, t given in (1), satisfy the equation (2) for some combination of the signs. This fits the well-known formula that if the values of $a_1, a_2, \ldots, a_n$ are $+1$ or -1, then

$$2 \sin\left(a_1 + \frac{a_1 a_2}{2} + \frac{a_1 a_2 a_3}{2^2} + \ldots + \frac{a_1 a_2 \ldots a_n}{2^n}\right)\frac{\pi}{4} =$$

$$= a_1 \sqrt{2 + a_2\sqrt{2 + \ldots + a_n\sqrt{2}}}.$$

1976/3. *A rectangular box can be completely filled with unit cubes. If one places as many cubes as possible, each with volume 2, in the box, with their edges parallel to the edges of the box, one can fill exactly 40% of the box. Determine the possible dimensions of the box.* $(\sqrt[3]{2} = 1.2599\ldots).$

Solution. Let a, b, c, denote the sides of the triangle and ε the side of the cube of volume 2 $(\varepsilon^3 = 2)$. From the approximation we get

(1) $$\frac{5}{4} = 1.25 < \varepsilon < 1.2857\ldots = \frac{9}{7}$$

When we fill exactly 40% (2/5-th) of the box with cubes of volume 2, then they can be pushed together to a rectangular block or we could insert more cubes of edge ε into the box.

Along the edge of the box of length a we can insert at most $\left[\dfrac{a}{\varepsilon}\right]$ many cubes, hence the volume of the filling is

$$2\left[\frac{a}{\varepsilon}\right]\left[\frac{b}{\varepsilon}\right]\left[\frac{c}{\varepsilon}\right] = abc \cdot \frac{2}{5},$$

hence

$$\frac{abc}{\left[\frac{a}{\varepsilon}\right]\left[\frac{b}{\varepsilon}\right]\left[\frac{c}{\varepsilon}\right]} = 5.$$

Let $t_n = \dfrac{n}{\left[\frac{n}{\varepsilon}\right]}$. Our problem is to determine the solutions of

(2) $$t_a t_b t_c = 5$$

Here is a table for the values of t_n:

n	2	3	4	5	6	7	8
t_n	2	$\dfrac{3}{2}$	$\dfrac{4}{3}$	$\dfrac{5}{3}$	$\dfrac{3}{2}$	$\dfrac{7}{5}$	$\dfrac{4}{3}$
	2,00	1.50	1.33	1.67	1.50	1.40	1.33

Now, it is convenient to make some upper and lower estimates for t_n. First — considering (1):

(3) $$t_n \geq \frac{n}{\frac{n}{\varepsilon}} = \varepsilon > \frac{5}{4},$$

on the other hand,

$$t_n = \varepsilon \frac{\frac{n}{\varepsilon}}{\left[\frac{n}{\varepsilon}\right]} < \varepsilon \frac{\left[\frac{n}{\varepsilon}\right]+1}{\left[\frac{n}{\varepsilon}\right]} = \varepsilon \left(1 + \frac{1}{\left[\frac{n}{\varepsilon}\right]}\right).$$

Now, if $n \geq 8$, $\left[\frac{n}{\varepsilon}\right] \geq \left[\frac{8}{\varepsilon}\right] = [6,34] = 6$, and so

(4) $$t_n < \varepsilon \left(1 + \frac{1}{6}\right) < \frac{9}{7} \cdot \frac{7}{6} = \frac{3}{2}.$$

For $n > 2$ in the table, $t_n \leq \frac{5}{3}$. Thus, if a, b and c are greater than 2,

$$t_a t_b t_c \leq \left(\frac{5}{3}\right)^3 = 4,62\ldots < 5,$$

contradicting (2); this also holds for $n \geq 8$; thus at least one of a, b and c equals 2. Let $a = 2$, then $t_a = 2$ and (2) becomes

(5) $$t_b t_c = \frac{5}{2}.$$

(3) implies that none of t_b and t_c can equal 2 and (4) implies that none of them can be greater then 7, because this gives

$$t_b t_c < \frac{3}{2} \cdot \frac{5}{3} = \frac{5}{2}, \quad \text{or} \quad t_b t_c < \left(\frac{3}{2}\right)^2 < \frac{5}{2}$$

contradicting (5). Hence

(6) $$3 \leq b, c \leq 7.$$

Since

$$abc = 5 \left[\frac{a}{\varepsilon}\right] \left[\frac{b}{\varepsilon}\right] \left[\frac{c}{\varepsilon}\right],$$

5 divides b or c, so (6) implies that $b = 5$ or $c = 5$; assume that $b = 5$, so $t_b = \frac{5}{3}$ then (5) implies $t_c = \frac{3}{2}$ and hence the possible values of c are $c = 3$ or $c = 6$.

The possible dimensions are $(2, 3, 5)$ or $(2, 5, 6)$. These rectangular blocks satisfy the conditions of the problem. The volume of the first one is 30 and one can place $1 \cdot 2 \cdot 3 = 6$ cubes into it, the volume of the second one is 60 and one can place $1 \cdot 3 \cdot 4 = 12$ cubes in it, the proportion of the volumes is 40% in both cases.

1976/4. *Determine the largest number which is the product of positive integers with sum* 1976.

First solution. As 1976 can be decomposed to a sum of integers only finitely many ways, there is a decomposition where the number of the summands is maximal. Let $a_1, a_2, \ldots, a_n$ denote the summands in this decomposition; thus $a_1 + a_2 + \ldots + a_n = 1976$. Let $M = a_1 a_2 \ldots a_n$.

There is no 1 among the a_i-s as omitting $a_i = 1$ and taking $a_k + 1$ instead of a_k we do not change the sum but increase the product as $1 \cdot a_k < a_k + 1$; thus $a_i \geq 2$.

We also show that $a_i < 5$: if, to the contrary, $a_i \geq 5$, then splitting a_i into the sum of $a_i' = 2$ and $a_i'' = a_i - 2$ we do not change the sum, but increase M, since

$$a_i' a_i'' = 2(a_i - 2) = 2a_i - 4 > a_i.$$

So, a_i equals 2, 3 or 4. 4 can be replaced by $2 + 2$, changing neither the sum, nor the product, so we may assume that there are only 2-s and 3-s in the sum.

There cannot be more then two 2-s in the sum, as replacing $2 + 2 + 2$ by $3 + 3$ the sum remains the same and the product becomes greater.

Our result is the following: if we split an arbitrary integer into a sum such that the product of the summands is maximal, there are at most two 2-s among the summands, the rest equal 3. Since $1976 = 3 \cdot 658 + 2$, $a_1 = 2$, $a_2 = a_3 = \ldots = a_{658} = 3$ and so $M = 2 \cdot 3^{658}$.

Second solution. Again, assume that from the decompositions of 1976 the product $M = a_1 a_2 \ldots a_n$ is maximal, and first assume that $n \geq 5$.

α) there cannot be two a_i-s with difference greater than 1 as if $a_k - a_j > 1$, then substituting $a_k' = a_k - 1$, $a_j' = a_j + 1$ the sum does not change, but the product increases:

$$a_k' a_j' = (a_k - 1)(a_j + 1) = a_k a_j + a_k - a_j - 1 > a_k a_j.$$

β) As for $k \neq 3$

$$3^k > k^3$$

(proof by induction), we can replace $a_i + a_i + a_i$ by a_i times 3 not changing the sum but increasing the product if $a_i \neq 3$.

Since $n \geq 5$, α) implies that one of the summands has to occur at least 3 times and by β) this has to equal 3. The other summand (that occurs at most twice) is 4 or 2. By the equality of their sum and product a 4 can be substituted by two 2-s, hence in the maximal decomposition there can occur only 3-s and at most two 2-s. As $1976 = 3 \cdot 658 + 2$, for $n \geq 5$ the maximal product is $3^{658} \cdot 2$.

If $n < 5$, by the A.M.–G.M. inequality the product cannot be larger than $\left(\dfrac{1976}{n} \right)^n$, but

$$\left(\frac{1976}{n} \right)^n < 2 \cdot 3^{658}$$

for every possible n.

Remarks. 1. In general, if a positive integer is of the form $3k+r$ $(r=0,$ 1, 2), then splitting it into a sum the product of its summands is maximum 3^k, $2^2 \cdot 3^{k-1}$, $2 \cdot 3^k$ for $r=0$, 1, 2, respectively.

2. If we want to split a number into the sum of k integers, then applying the arguments of our methods in case $n=kt+r$ $(0 \le r < k)$, the maximal product is

$$(t+1)^r t^{k-r}.$$

3. The most general form of the problem is the following: Split the real number n into a sum such that the product of the summands is maximal. Here, the only problem is to determine the number of the summands, because the A.M.– G.M. inequality tells us that they have to be equal. Using calculus one can show that in case of a maximal product the number of summands equals $k = \left[\dfrac{n}{e} \right]$, ($e$ the Euler-constant $e = 2.7182818\ldots$), or $k+1$ and so the maximal value is

$$\left(\frac{n}{k} \right)^k \quad \text{or} \quad \left(\frac{n}{k+1} \right)^{k+1}.$$

As $\dfrac{n}{k}$, or $\dfrac{n}{k+1}$ is approximately e, one can say that in case of a maximal product the summands equal approximately e. This is the deeper reason of our results as 3 is the integer closest to e.

1976/5. *Consider the following system of p equations with $q=2p$ unknowns:*

$$a_{11}x_1 + a_{12}x_2 + \ldots + a_{1q}x_q = 0,$$
$$a_{21}x_1 + a_{22}x_2 + \ldots + a_{2q}x_q = 0,$$
$$\cdots\cdots\cdots\cdots\cdots\cdots\cdots\cdots\cdots$$
$$a_{p1}x_1 + a_{p2}x_2 + \ldots + a_{pq}x_q = 0,$$

where $a_{ij} \in \{0, 1, -1\}$ $(i=1, 2, \ldots, p, \ j=1, 2, \ldots, q)$.

Prove that the system of equations has a solution $x_1, x_2, \ldots, x_q$ satisfying the following properties:

a) *$x_1, x_2, \ldots, x_q$ are integers;*

b) *not all x_j are 0 $(j=1, 2, \ldots, q)$;*

c) *$|x_j| \le q$ $(j=1, 2, \ldots, q)$.*

Solution. Let us suppose that substituting the two different q-tuples of integers

$$(u_1, u_2, \ldots, u_q) \quad \text{and} \quad (v_1, v_2, \ldots, v_q)$$

where $|u_j| \le p$, $|v_j| \le p$ $(j=1, 2, \ldots, q)$ into the i-th equation we get the same values y_i, $(i=1, 2, \ldots, p)$. Then

$$(u_1 - v_1, u_2 - v_2, \ldots, u_q - v_q)$$

satisfy the equations and the conditions a), b), c): a) and b) directly follow from the choice of the u_i-s and v_i-s and c) holds because $|u_i - v_i| \le |u_i| + |v_i| \le 2p = q$.

Therefore to solve the problem we only have to show the existence of the u_i-s and v_i-s.

In such a q-tuple $(u_1, u_2, \ldots, u_q)$ as $-p \leq u_i \leq p$, there are $2p+1$ choices for u_i so there are

$$(2p+1)^q = (2p+1)^{2p} = (4p^2+2p+1)^p.$$

q-tuples. By our definition

$$y_i = a_{i_1} u_1 + a_{i_2} u_2 + \ldots + a_{i_q} u_q \qquad (i = 1, 2, \ldots, p),$$

hence

$$|y_i| \leq |a_{i_1}||u_1| + |a_{i_2}||u_2| + \ldots + |a_{i_q}||u_q| \leq$$
$$\leq |u_1| + |u_2| + \ldots + |u_q| \leq pq,$$

so y_i can attain at most $2pq+1$ different values, so the number of the possible p-tuples for $(y_1, y_2, \ldots, y_p)$ is at most

$$(2pq+1)^p = (4p^2+1)^p.$$

But, as

$$(4p^2+4p+1)^p > (4p^2+1)^p,$$

There are two distinct q-tuples producing the same results; and this is what we wanted to prove.

1976/6. *The sequence $u_0, u_1, \ldots$ is defined as follows: $u_0 = 2$, $u_1 = \dfrac{5}{2}$, $u_{n+1} = u_n(u_{n-1}^2 - 2) - u_1$ $(n = 1, 2, \ldots)$. Prove that*

$$u_n = 2^{\frac{2^n - (-1)^n}{3}} \qquad (n = 1, 2, \ldots).$$

Solution. Construct the first few elements of the series:

$$u_0 = 1 + \frac{1}{1}, \ u_1 = 2 + \frac{1}{2}, \ u_2 = 2 + \frac{1}{2}, \ u_3 = 8 + \frac{1}{8}, \ u_4 = 32 + \frac{1}{32}, \ u_5 = 2048 + \frac{1}{2048}.$$

They can be rewritten in the form

$$u_0 = 2^0 + \frac{1}{2^0}, \quad u_1 = 2^1 + \frac{1}{2^1}, \quad u_2 = 2^1 + \frac{1}{2^1},$$
$$u_3 = 2^3 + \frac{1}{2^3}, \quad u_4 = 2^5 + \frac{1}{2^5}, \quad u_5 = 2^{11} + \frac{1}{2^{11}}.$$

So we formulate the following conjecture:

(1) $$u_n = 2^{f(n)} + 2^{-f(n)}$$

where

(2) $$f(n) = \frac{2^n - (-1)^n}{3}.$$

The conjecture holds for $n = 0, 1, \ldots, 5$. We prove the statement by induction. Assume that (1) and (2) holds for until some integer n. By the definition of the sequence

$$u_{n+1} = \left(2^{f(n)} + 2^{-f(n)}\right)\left(2^{2f(n-1)} + 2^{-2f(n-1)}\right) - \frac{5}{2} =$$

$$(3) \qquad = 2^{f(n)+2f(n-1)} + 2^{-f(n)+2f(n-1)} + 2^{f(n)-2f(n-1)} + 2^{-f(n)-2f(n-1)} - \frac{5}{2}.$$

Examining the exponents according to (2) we have:

$$f(n) + 2f(n-1) = \frac{1}{3}\left(2^n - (-1)^n + 2^n - (-1)^{n-1} \cdot 2\right) =$$

$$= \frac{1}{3}\left(2^{n+1} - (-1)^{n+1}\right) = f(n+1),$$

$$f(n) - 2f(n-1) = \frac{1}{3}\left(2^n - (-1)^n - 2^n + (-1)^{n-1} \cdot 2\right) =$$

$$= \frac{1}{3}\left(-(-1)^n + (-1)^{n-1} \cdot 2\right) = \frac{1}{3}\left((-1)^{n+1} + 2(-1)^{n+1}\right) = (-1)^{n+1}.$$

Using this for (3) we get:

$$u_{n+1} = 2^{f(n+1)} + 2^{-f(n+1)} + 2 + 2^{-1} - \frac{5}{2} = 2^{f(n+1)} + 2^{-f(n+1)},$$

so we proved our conjecture.

$f(n)$ is always an integer, because 3 divides $2^n - (-1)^n$; but $2^{-f(n)} < 1$, hence

$$[u_n] = \left[2^{f(n)} + 2^{-f(n)}\right] = 2^{f(n)} = 2^{\frac{2^n-(-1)^n}{3}}.$$

1977.

1977/1. *Construct equilateral triangles ABK, BCL, CDM, DAN on the inside of the square $ABCD$. Show that the midpoints of KL, LM, MN, NK and the midpoints of AK, BK, BL, CL, CM, DM, DN, AN form a regular dodecahedron.*

Solution. We can observe lots of symmetries of our object; the combination of them can lead to the proof in several ways. We choose one of these approaches (see *Figure 1977/1.1*).

The equilateral triangle ABK is the rotated image of the triangle BCL by $90°$, hence BK and CL are perpendicular. Moreover BK lies on the perpendicular bisector of CL intersecting it in X, so X is the midpoint of CL. Similarly, AK and DN intersect in Y, the midpoint of DN. Let Z denote the midpoint of LK, Z is on the BD diagonal. Let O denote the centre of the square. L and M are symmetric to the diagonal AC, so the midpoint U of LM, lies on AC and LM is orthogonal to AC.

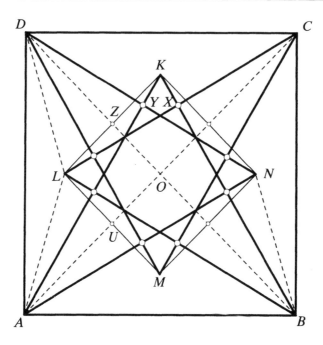

Figure 77/1.1

To prove our theorem it is enough to show that ZOY and XOY are congruent equilateral triangles (the central triangles of the dodecahedron), because the other two central triangles are the reflections of one of these triangles to the diagonals of the square. The quadrangle $KLMN$ is a square as the rotation by $90°$ about O fixes it, thus $ZO = UO$.

If we join a vertex of a triangle inside the square (e.g: L) with the nearest vertices (A and D), then the base angles of the isosceles triangle LAD are $15°$, because the base angles of the isosceles triangle ABL are $75°$.

In the congruent triangles ACL and BDN we have $AL = BN$, and XO is the median parallel to AL in the first, YO is the median parallel to BN in the second triangle, so $XO = YO$. As AL and BN include angles of $15°$—$15°$ with the direction of AD, the vertex angle of the isosceles triangle XOY is $30°$. As $\angle NBD = 30°$, the similarly oriented angle $\angle YOZ$ is $30°$, as well. $ZO = LU = AL/2$, because one of the angles of the right triangle, AUL is $30°$ and as $YO = AL/2$, ZOY and XOY are congruent equilateral triangles.

1977/2. *In a finite sequence of real numbers the sum of any seven consecutive terms is negative, and the sum of any eleven consecutive terms is positive.*

Determine the maximum number of terms in the sequence.

First solution. We show that such a sequence cannot have 17 or more terms, but there is a sequence with 16 terms. Let us assume that the sequence has 17 terms: $a_1, a_2, \ldots, a_{17}$; order them by the following scheme:

$$
\begin{array}{ccccc}
a_1 & a_2 & a_3 & \ldots & a_7 \\
a_2 & a_3 & a_4 & \ldots & a_8 \\
a_3 & a_4 & a_5 & \ldots & a_9 \\
\ldots & \ldots & \ldots & \ldots & \ldots \\
a_{11} & a_{12} & a_{13} & \ldots & a_{17}.
\end{array}
$$

The sum of the terms is positive in every column and negative in every row. If we sum up all the numbers of the table by the rows, the sum is positive, if by the columns, it is negative, a contradiction. So we can have at most 16 terms in the sequence. That is possible by the following example:

$$8, \ 8, \ -21, \ 8, \ 8, \ 8, \ -21, \ 8, \ 8, \ -21, \ 8, \ 8, \ 8, \ -21, \ 8, \ 8.$$

Thus the maximum number of the terms is 16.

Second solution. The advantage of the following solution is that it enables us to construct infinitely many examples with maximal number of terms.

Let s_i denote the sum of the first i elements, $s_0 = 0$. The conditions of the problem say that

(1) $\qquad s_{i+7} - s_i < 0, \quad \text{i.e.} \quad s_i > s_{i+7} \quad (i = 0, 1, 2, \ldots)$

and

(2) $\qquad s_i - s_{i-11} > 0, \quad \text{i.e.} \quad s_i > s_{i-11} \quad (i = 11, 12, \ldots)$

If the sequence has 17 elements, then iterating the two inequalities we get:

$$\downarrow$$

(3) $\qquad s_7 > s_{14} > s_3 > s_{10} > s_{17} > s_6 > s_{13} > s_2 > s_9 > s_{16} >$

$\qquad > s_5 > s_{12} > s_1 > s_8 > s_{15} > s_4 > s_{11} > s_0 = 0,$

that is a contradiction, because $s_7 < 0$.

Now, let us assume that the sequence has only 16 terms and start the chain of inequalities at s_6 (denoted by an arrow in (3)), and continue from s_0 as follows:

(4) $\qquad s_0 > s_7 > s_{14} > s_3 > s_{10}.$

Now the chain is interrupted as none of (1) and (2) apply. Give arbitrary values to the s_i- s satisfying (3) and (4), e.g: $12 > 11 > 10 > \ldots$, and construct the table:

s_1	s_2	s_3	s_4	s_5	s_6	s_7	s_8	s_9	s_{10}	s_{11}	s_{12}	s_{13}	s_{14}	s_{15}	s_{16}
5	10	−3	2	7	12	−1	4	9	−4	1	6	11	−2	3	8

Since $a_i = s_i - s_{i-1}$, the table produces:

$$5, \ 5, \ -13, \ 5, \ 5, \ 5, \ -13, \ 5, \ 5, \ -13, \ 5, \ 5, \ 5, \ -13, \ 5, \ 5.$$

As we have infinitely many choices for the s_i-s, we can construct infinitely many sequences with 16 terms satisfying the conditions of he problem.

Third solution. We show that the assumption that the sequence has 17 terms leads to a contradiction.

Choose arbitrary $11 - 7 = 4$ consecutive terms in the sequences. As the sequence has 17 terms, these 4 terms are the first or last 4 terms of an 11 tuple. As the sum of the 11 terms is positive and the sum of the complementary 7 terms is negative, the sum of the four terms has to be positive; thus the sum of any 4 consecutive terms is positive.

Now, choose arbitrary $7 - 4 = 3$ consecutive terms. There is a 7 tuple in the sequence such that these 3 terms are the first or last 3 terms of a a 7 tuple . The sum of the 7 terms is negative, the sum of the complementary 4 terms is positive, so the sum of the 3 terms has to be negative.

Now, choose arbitrary $4 - 3 = 1$ term. This has to be positive as this is the complement of 3 terms in a 4-tuple where the sum of the 3 terms is negative and the sum of the 4-tuple is positive. Thus every term has to be positive so there is no sequence satisfying the conditions of length 17.

As shown in the previous solutions, there are sequences of length 16 satisfying the conditions.

Remarks. 1. With the method of the first solution it can be shown that if in a sequence the sum of any consecutive n terms is negative and the sum of any consecutive p terms is positive then the sequence has at most $n + p - 1$ terms.

2. The method of the third solution reminds us to the Euclidean algorithm. Indeed, if the sum of any consecutive n terms is negative and the sum of any consecutive p terms is positive, then the sequence cannot have more than $n + p - - \gcd(n, p)$ terms, where gcd stands for the greatest common divisor. If — as in our problem — n and p are coprime, then the sequence has at most $n + p - 1$ terms.

1977/3. *Given an integer $n > 2$, let V_n be the set of integers $1 + kn$ for k a positive integer. A number $m \in V_n$ is called indecomposable if there is no number $p, q \in V_n$ such that $pq = m$.*

Prove that there is a number r in V_n which can be expressed as the product of indecomposable members of V_n in more than one way (decompositions which differ solely in the order of factors are not regarded as different).

First solution. First, we collect a few simple observations about the numbers in V_n:

α) 1 is not in V_n.

β) The products of two members of V_n is in V_n: if $k_1 n + 1 \in V_n$ and $k_2 n + 1 \in V_n$, then

$$(k_1 n + 1)(k_2 n + 1) = (k_1 k_2 n + k_1 + k_2)n + 1 \in V_n.$$

γ) the product of $k_1 n - 1$ and $k_2 n - 1$ is also in V_n:

$$(k_1 n - 1)(k_2 n - 1) = (k_1 k_2 n - k_1 - k_2)n + 1. \quad (n > 2).$$

The beginning of the list of the members of V_n in increasing order is:

$$n+1, \ 2n+1, \ 3n+1, \ \ldots, \ (n+1)n+1, \ (n+2)n+1 = (n+1)^2.$$

We have to find a number in V_n that splits into the product of indecomposable numbers in two different ways in V_n.

Let $a = n - 1$, $b = 2n - 1$. By β) and γ) we have that a^2, b^2, ab, $(ab)^2$ are in V_n:

$$ab \cdot ab = a^2 \cdot b^2$$

are two decompositions of $(ab)^2$. These two decompositions are different as e.g. $a^2 = ab$ implies $a = b$ that is a contradiction.

We show that a^2, b^2, ab are indecomposable in V_n.

a^2 is indecomposable, because it is smaller than , $(n+1)^2$, the smallest decomposable number in V_n.

If $b^2 = (2n-1)^2$ is decomposable, then one of its factors has to be $n+1$, because $(2n+1)^2 > (2n-1)^2$ and $2n+1$ is the second smallest term of V_n. Hence in case b^2 were decomposable, then

$$b^2 = (2n-1)^2 = (n+1)(cn+1) \quad (c \text{ positive integer})$$

possible, and so

$$n = \frac{c+5}{4-c}.$$

This is an integer only if $c=3$ and $n=8$. So if $n \neq 8$, b^2 is indecomposable.

In a decomposition of $ab = (n-1)(2n-1)$ both factors cannot be greater than $(n+1)$ as $(2n+1)^2 > (n-1)(2n-1)$, hence if ab can be decomposed, it is done in the following form:

$$(n-1)(2n-1) = (n+1)(dn+1)$$

where d is a positive integer. Hence

$$n = \frac{d+4}{2-d},$$

that holds in case of $d=1$ and $n=5$. Thus ab is decomposable if $n \neq 5$.

We solved the problem if $n \neq 5$ or $n \neq 8$. We show that the statement holds in these latter cases, as well.

In case $n = 5$, $3 \cdot 5 + 1 = 16$, $15 \cdot 5 + 1 = 76$, $72 \cdot 5 + 1 = 361$ are indecomposable numbers in V_5 because they do not have two divisors of the form $5k + 1$, but

$$76^2 = 76 \cdot 76 = 16 \cdot 361,$$

so $76^2 \in V_5$ is decomposed in V_5 in two different ways.

If $n = 8$, $8 \cdot 8 + 1 = 65$, $3 \cdot 8 + 1 = 25$, $21 \cdot 8 + 1 = 169$ are indecomposable numbers in V_8 because they do not have two divisors of the form $8k + 1$, but

$$65^2 = 65 \cdot 65 = 25 \cdot 169,$$

and so $65^2 \in V_8$ is decomposed in two different ways.

Second solution. The following solution refers to the origins of the problem, but it uses a deep theorem of number theory, namely Dirichlet's theorem: In the arithmetic progression $an + b$ where a, b are coprime integers there are infinitely many primes.

We shall utilize the theorem through the following statement: there are three primes of the form $nk - 1$:

$$p_1 = nq - 1, \quad p_2 = nr - 1, \quad p_3 = ns - 1,$$

where q, r, s are positive integers. None of these numbers is in V_n as, for example,

$$nq - 1 = nk + 1$$

implies $n(q - k) = 2$, and it is impossible since $n > 2$.

$p_1 p_2$, $p_2 p_3$, $p_3 p_1$ and p_3^2 are in V_n, but they are indecomposable in V_n since they have no prime divisors in V_n. $p_1 p_2 p_3^2 \in V_n$ can be decomposed in V_n in at least two ways:

$$(p_1 p_3)(p_2 p_3) = (p_1 p_2)p_3^2.$$

The two decompositions are different since $p_1 p_3 = p_3^2$ would imply $p_1 = p_3$. Hence we proved our statement.

Remarks. 1. The condition $n > 2$ is necessary, because in case of $n = 2$, V_n is the set of odd primes; here, the indecomposable numbers are the primes and if an odd number is a product of two primes, it can be done in a unique way.

2. Using Dirichlet's theorem is a very strong theorem in the second solution, but the jury accepted it as a full solution. The third solution will use a simple special case of the theorem.

Third solution. First, we prove three lemmas:

Lemma 1. There are infinitely many primes that are not of the form $kn + 1$ (n, k positive integers, $n > 1$ a fix integer).

Let us assume to the contrary that there are only finitely many primes of this kind, $t_1, t_2, \ldots, t_s$ and let $U = nt_1 t_2 \ldots t_s - 1$. None of the t_i-s divide U, hence all its prime factors are of the form $kn+1$; the product of these primes is of the form $kn+1$, but U is not. Thus there are infinitely many primes not of the form $kn+1$.

Lemma 2. There is an integer a, such that there are infinitely many primes of the form $kn+a$ (k positive integer).

According to Lemma 1. there are infinitely many primes of the form $kn+r$, where $1<r<n$. The possible values of r are 2, 3, $\ldots$, $n-1$, hence there is $r=a$, such that there are infinitely many primes of the form $kn+a$. Let P denote the set of these primes.

Lemma 3. There is a positive exponent $\alpha \leq n$ such that a^{α} is of the form $kn+1$, where a is the positive integer defined in Lemma 2.

First, observe that there are two powers in the sequence $a, a^2, \ldots, a^{n+1}$ that give the same residue mod n, e.g., a^{α_1} and a^{α_2} ($\alpha_2 > \alpha_1$). Then n divides $a^{\alpha_2} - a^{\alpha_1} = a^{\alpha_1}(a^{\alpha_2-\alpha_1} - 1)$. a and n are coprime as the elements of P are divisible by the greatest common divisor of a and n, so it has to equal 1. So n divides $a^{\alpha_2-\alpha_1} - 1$, hence for $\alpha_2 - \alpha_1 = \alpha$, $a^{\alpha} - 1 = kn$ (k positive integer), thus

(1) $a^{\alpha} = kn+1.$

α is greater than 1, because $\alpha = 1$ implies $a = kn+1$ that is impossible since $a < n$, thus $\alpha \geq 2$. Let u denote the smallest of the α-s satisfying (1) ($u \geq 2$); $p_1, p_2, \ldots, p_u$ be arbitrary primes in P. $p = p_1 p_2 \ldots p_u$ is of the form $An + a^u$, where A is a positive integer. By the choice of u, a^u is of the form $Bn+1$ (B positive integer), hence $p = Cn+1$ (C positive integer), so p is in V_n.

p is indecomposable in V_n, because in the other case the product of a few (let's say s) of the primes $p_1, p_2, \ldots, p_u$, was in V_n, i.e. of the form $kn+1$. This product is of the form $Dn + a^s$ (D positive integer), that is a^s gives residue 1 mod n contradicting the definition of u. Thus the product of any u elements from P is in V_n and it is indecomposable.

Finally, let $p_1, p_2, \ldots, p_u$; $q_1, q_2, \ldots, q_u$ be primes in P. Their product

$$p_1 p_2 \ldots p_u q_1 q_2 \ldots q_u = (p_1 p_2 \ldots p_u)(q_1 q_2 \ldots q_u) =$$
$$= (q_1 p_2 p_3 \ldots p_u)(p_1 q_2 q_3 \ldots q_u)$$

factors into the product of indecomposables in at least two ways.

1977/4. *Define*

(1) $f(x) = 1 - a \cos x - b \sin x - A \cos 2x - B \sin 2x$

where a, b, A and B are real constants. Suppose that $f(x) \geq 0$ for all real x. Prove that

(2) $a^2 + b^2 \leq 2, \qquad A^2 + B^2 \leq 1.$

Solution. If $a = b = A = B = 0$, then (2) clearly holds. Assume that at least one of a, b, and A, B is non-zero. Then

$$f(x) = 1 - \sqrt{a^2+b^2} \left(\frac{a}{\sqrt{a^2+b^2}} \cos x + \frac{b}{\sqrt{a^2+b^2}} \sin x \right) -$$

$$- \sqrt{A^2 + B^2} \left(\frac{A}{\sqrt{A^2 + B^2}} \cos 2x + \frac{B}{\sqrt{A^2 + B^2}} \sin 2x \right).$$

Let α and 2β denote the angles of the unit vectors $\mathbf{e} \left(\dfrac{a}{\sqrt{a^2 + b^2}}, \dfrac{b}{\sqrt{a^2 + b^2}} \right)$ and

$\mathbf{E} \left(\dfrac{A}{\sqrt{A^2 + B^2}}, \dfrac{B}{\sqrt{A^2 + B^2}} \right)$ respectively. Using the above notations

$$f(x) = 1 - \sqrt{a^2 + b^2} \cos(x - \alpha) - \sqrt{A^2 + B^2} \cos 2(x - \beta),$$

and this holds even if $a^2 + b^2 = 0$, or $A^2 + B^2 = 0$. As $f(x) \geq 0$,

$$f(\beta) + f(\pi + \beta) = 1 - \sqrt{a^2 + b^2} \cos(\beta - \alpha) - \sqrt{A^2 + B^2} +$$

$$+ 1 + \sqrt{a^2 + b^2} \cos(\beta - \alpha) - \sqrt{A^2 + B^2} = 2(1 - \sqrt{A^2 + B^2}) \geq 0,$$

Hence

$$A^2 + B^2 \leq 1.$$

Moreover,

$$f\left(\alpha + \frac{\pi}{4}\right) + f\left(\alpha - \frac{\pi}{4}\right) = 1 - \sqrt{a^2 + b^2} \cdot \frac{\sqrt{2}}{2} + \sqrt{A^2 + B^2} \sin 2(\alpha - \beta) +$$

$$+ 1 - \sqrt{a^2 + b^2} \frac{\sqrt{2}}{2} - \sqrt{A^2 + B^2} \sin 2(\alpha - \beta) = 2 - \sqrt{2}\sqrt{a^2 + b^2} \geq 0,$$

and $$a^2 + b^2 \leq 2,$$

and so (2) is proved.

Remark. The converse of the statement is not true, (2) does not imply $f(x) \geq 0$; e.g. in case $a = \sqrt{2}$, $b = A = B = 0$ $a^2 + b^2 \leq 2$, $A^2 + B^2 \leq 1$, but $f(0) = 1 - \sqrt{2} < 0$.

1977/5. *Let a and b be positive integers. When $(a^2 + b^2)$ is divided by $(a + b)$, the quotient is q, the remainder r.*

Find all pairs (a, b), such that

(1) $$q^2 + r = 1977.$$

Solution. By the conditions

(2) $$a^2 + b^2 = (a + b)q + r, \quad \text{where} \quad 0 \leq r < a + b,$$

and so

$$q + \frac{r}{a + b} = \frac{a^2 + b^2}{a + b}.$$

$r < a + b$ implies $\dfrac{r}{a + b} < 1$, and using the A.M.–G.M. inequality:

$$q + 1 > \frac{a^2 + b^2}{a + b} \geq \frac{a + b}{2},$$

giving

(3) $$2(q+1) > a+b.$$

(1) implies

$$q^2 \leq 1977, \quad \text{thus} \quad q \leq 44.$$

By (3)

$$r < a+b < 2 \cdot 45 = 90.$$

On the other hand, using (1) we get:

$$q^2 = 1977 - r > 1977 - 90 = 1887,$$

that is

$$1887 < q^2 < 1977.$$

As only $1936 = 44^2$ is a square, $q = 44$ and so $r = 1977 - 1936 = 41$. Substituting to (2), we get

$$a^2 + b^2 = 44(a+b) + 41,$$
$$(a-22)^2 + (b-22)^2 = 1009.$$

None of the squares on the l.h.s can be smaller than the half of 1009 so they are at least 505, hence the possible values are the squares between 505 and 1009:

$$529, \quad 576, \quad 625, \quad 676, \quad 729, \quad 784, \quad 841, \quad 900, \quad 961.$$

Considering the last digit, we get that only 784 gives a square difference with 1009, with $15^2 = 225$, so

$$\{|a-22|, |b-22|\} = \{28, 15\},$$

hence the possible (a, b) pairs are:

$$(7, 50), \quad (37, 50), \quad (50, 7), \quad (50, 37),$$

and these values satisfy the conditions of the problem.

1977/6. *The function f is defined on the set of positive integers and its values are positive integers. Let us assume that*

(1) $$f(n+1) > f(f(n))$$

for every positive n.

Prove that for every positive n

$$f(n) = n.$$

First solution. We prove the statement by induction. First, we show that $f(1) = 1$. There has to be a k_0 such that $f(k_0) = 1$. If not, let us consider the following sequence

$$k_1 = f(k_0), \quad k_2 = f(k_1 - 1), \quad \ldots, \quad k_i = f(k_{i-1} - 1), \quad \ldots$$

As by our assumptions $f(k_{i-1} - 1) = k_i \neq 1$, $k_i > 1$, hence $k_i - 1$ is a positive integer, and so this sequence exists. By (1) we get $f(k_i) > f(f(k_i - 1)) = f(k_{i+1})$, hence there exists an infinite monotone decreasing sequence of integers

$$f(k_0) > f(k_1) > \ldots > f(k_i) > f(k_{i+1}) > \ldots$$

which is impossible.

Thus there is a positive integer k_0 such that $f(k_0) = 1$. But since if $k_0 > 1$, $k_0 - 1 \geq 1$ the inequality in (1) would imply

$$f(f(k_0 - 1)) < f(k_0) = 1$$

which is impossible, as f cannot attain a value smaller than 1. Therefore

$$f(1) = 1.$$

Now, assume that the statement is true for some positive integer k for every function f satisfying the conditions of the problem, i.e. $f(k) = k$. We have to prove that

$$f(k + 1) = k + 1.$$

Consider the function

$$\varphi(n) = f(n + 1) - 1.$$

We show that $\varphi(n)$ satisfies the conditions of the problem. (1) implies $f(n + 1) > 1$, and so $f(n + 1) - 1$ is a positive integer. Hence — because of (1):

$$\varphi(\varphi(n)) = \varphi(f(n + 1) - 1) = f(f(n + 1)) - 1 < f(n + 2) - 1 = \varphi(n + 1),$$

that is, $\varphi(n)$ satisfies (1) and by the assumption of the induction

$$\varphi(k) = k.$$

So

$$\varphi(k) = f(k + 1) - 1 = k, \qquad f(k + 1) = k + 1,$$

And this is what we wanted to prove.

Second solution. We prove that f is strictly increasing. Let us divide the set of integers into two disjoint subsets K and M.

Let K be the set of positive integers k such that

(2) $$f(k) < k;$$

and M be the set of positive integers n such that

(3) $$f(n) \geq n.$$

First we show that K is the empty set. Indeed, if K had an element, choose m in K such that $f(m)$ is the smallest. By the definition of K

(4) $$f(m) < m, \quad \text{that is} \quad f(m) \leq m - 1.$$

$m - 1$ is positive as it is not smaller than a value of the function, and so, $f(m - 1)$ exists. (1) and (4) imply that

(5) $$m > f(m) > f(f(m - 1)).$$

(5) implies that $f(f(m - 1))$ is smaller than $f(m)$. But f cannot attain values smaller the $f(m)$ in K, thus $f(m - 1) \in M$. Hence, applying (3) and (5) we

get:

(6) $$m > f(m) > f(f(m-1)) \geq f(m-1).$$

Similarly, $m-1$ is not in K either, because $f(m-1) < f(m)$ and the values of f at the elements of K cannot be less than $f(m)$, hence $m-1 \in M$ and by (3)

$$f(m-1) \geq m-1.$$

Combining with (4) we get

$$f(m-1) \geq f(m).$$

This contradicts (6), so $m-1$ cannot be the element of K, so (3) holds for every positive integer n. Hence applying (1) and (3)

$$f(n+1) > f(f(n)) \geq f(n),$$

that is

$$f(n+1) > f(n)$$

follows, hence f is strictly increasing.

Now, it follows that in (3) equality holds because if for some n $f(n) > n$, it means $f(n) \geq n+1$, but by the monotonity

$$f(f(n)) \geq f(n+1)$$

holds, contradicting (1). Thus for every n

$$f(n) = n,$$

and this is what we wanted to prove.

1978.

1978/1. *For m and n positive integers, $n > m > 1$, the last three decimal digits of 1978^m is the same as the last three decimal digits of 1978^n. Find m and n such that $m+n$ has the least possible value.*

Solution. 1978^n and 1978^m has the same last three digits if and only if 1000 divides their difference:

$$1978^n - 1978^m = 1978^m(1978^{n-m} - 1)$$

As $1000 = 8 \cdot 125$, we have to examine divisibility by 8 and 125.

The second factor is odd and only the first power of 2 divides 1978. Hence m has to be at least 3, $m \geq 3$.

Only the factor $1978^{n-m} - 1$ can be divisible by 125. As the last digits of the powers of 1978 are 8, 4, 2, 6, 8, 4, ..., their period is 4. Obviously, 5 divides $1978^{n-m} - 1$ only in case the last digit is 6, that is if $n - m = 4k$. Since the remainder of 1978 mod 125 is -22 and the remainder of $(-22)^4$ is 6, 125 divides $1978^{4k} - 1$ if and only if 125 divides $6^k - 1 = (1+5)^k - 1$.

Applying the binomial theorem:

$$(1+5)^k - 1 = 5k + 5^2\frac{k(k-1)}{2} + \dots$$

the other terms are divisible by $5^3 = 125$ hence we need to examine

$$5k + 25\frac{k(k-1)}{2} = \frac{5k}{2}(5k-3).$$

$5k - 3$ is relatively prime to 125 and $5k$ is divisible by 125, if 25 divides k, and so k is at least 25, that is $n - m = 4k \geq 100$. Since $m \geq 3$, $n + m = n - m + 2m \geq \geq 106$. Thus the minimal value of $n + m$ is 106, and $m = 3$, $n = 103$.

1978/2. *P is a point inside a sphere. Three mutually perpendicular rays from P intersect the sphere at points U, V and W. Let Q denote the vertex diagonally opposite P in the parallelepiped determined by PU, PV, PW. Find the locus of Q for all possible sets of such rays from P.*

Solution. Let R denote the radius, O the centre of the sphere, $OP = p$. We show that the locus is a sphere with centre O and radius $\sqrt{3R^2 - 2p^2}$.

Let $\mathbf{a}$, $\mathbf{b}$, $\mathbf{c}$, $\mathbf{p}$, $\mathbf{q}$ denote the vectors pointing from O to A, B, C, P, Q, respectively. As $\mathbf{q} - \mathbf{p}$ is the diagonal vector of the parallelepiped,

$$\mathbf{q} - \mathbf{p} = (\mathbf{a} - \mathbf{p}) + (\mathbf{b} - \mathbf{p}) + (\mathbf{c} - \mathbf{p}),$$

Since $(\mathbf{a} - \mathbf{p})$, $(\mathbf{b} - \mathbf{p})$ and $(\mathbf{c} - \mathbf{p})$ are mutually perpendicular vectors and $\mathbf{a}^2 = = \mathbf{b}^2 = \mathbf{c}^2 = R^2$,

$$\mathbf{q}^2 = \left(\mathbf{p} + (\mathbf{a} - \mathbf{p}) + (\mathbf{b} - \mathbf{p}) + (\mathbf{c} - \mathbf{p})\right)^2 =$$

$$= \mathbf{p}^2 + (\mathbf{a} - \mathbf{p})^2 + (\mathbf{b} - \mathbf{p})^2 + (\mathbf{c} - \mathbf{p})^2 + 2\mathbf{p}(\mathbf{a} + \mathbf{b} + \mathbf{c} - 3\mathbf{p}) = 3R^2 - 2p^2,$$

thus we showed that the points Q are on the designated sphere.

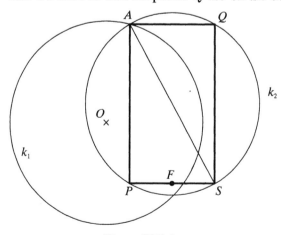

Figure 78/2.1a

We have to show that any point of the sphere can be the Q vertex of an appropriate parallelepiped. We can choose the parallelepiped in a special way.

Let the plane spanned by P, Q and O intersect the original sphere in the circle k_1, and the sphere with diameter PQ in the circle k_2. One of the points of intersection of k_1 and k_2 is A. Let S complete the right triangle PAQ to the $PAQS$ rectangle (see *Figures 1978/2.1a, b*). Now, let us build a square base prism such that this rectangle is a diagonal section of it, with base square $PBSC$, and on the cover the points A and Q are opposite vertices. Let F denote the midpoint of PS. As the diagonal BC is orthogonal to the plane of the diagonal intersection, i.e. to the plane of k_1, the triangle OFB is a right triangle.

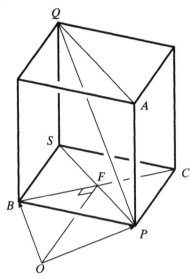

In order to prove our statement it is enough to show that B and C are on the sphere; Since O lies on the perpendicular bisector plane of BC, it is enough to show that $OB = R$.

We shall use that for an arbitrary point M in the space, the distances from the vertices of the rectangle $XYZU$ satisfy $MX^2 + MZ^2 = MY^2 + MU^2$ (see Remark 1.). We shall denote the vectors by bold face lower case letters corresponding to the points labelled by capital letters with O as the vector origin. Applying the theorem to the rectangle $APSQ$ and the point O, we obtain

$$\mathbf{s}^2 = \mathbf{p}^2 + \mathbf{q}^2 - \mathbf{a}^2 =$$

$$= p^2 + 3R^2 - 2p^2 - R^2 = 2R^2 - \mathbf{p}^2.$$

Figure 78/2.1b

In the square $PBSC$ we have $FB = \dfrac{PS}{2} = \dfrac{|\mathbf{s}-\mathbf{p}|}{2}$, but $|\overrightarrow{OF}| = \left|\dfrac{\mathbf{s}+\mathbf{p}}{2}\right|$, hence from the right triangle OFB

$$OB^2 = OF^2 + FB^2 = \left(\frac{\mathbf{s}+\mathbf{p}}{2}\right)^2 + \left(\frac{\mathbf{s}-\mathbf{p}}{2}\right)^2 = \frac{\mathbf{s}^2+\mathbf{p}^2}{2} = \frac{2R^2-p^2+p^2}{2} = R^2,$$

B and C are on the given sphere and the square base prism satisfies the conditions of the problem.

Remarks. 1. The rectangle theorem we used above can be easily proved by vector methods; let $\mathbf{x}$, $\mathbf{x}+\mathbf{a}$, $\mathbf{x}+\mathbf{a}+\mathbf{b}$, $\mathbf{x}+\mathbf{b}$ represent the vectors corresponding to the vertices of the rectangle $XYZU$, choosing M as a vector origin, where $\mathbf{a}$, $\mathbf{b}$ denote the side vectors of the rectangle (see *Figure 1978/2.2*). From $\mathbf{ab} = 0$ it follows

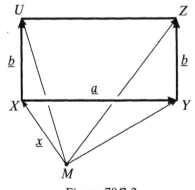

Figure 78/2.2

$$\mathbf{x}^2 + (\mathbf{x}+\mathbf{a}+\mathbf{b})^2 = 2\mathbf{x}^2 + \mathbf{a}^2 + \mathbf{b}^2 + 2\mathbf{ax} + 2\mathbf{bx},$$

$$(\mathbf{x}+\mathbf{a})^2 + (\mathbf{x}+\mathbf{b})^2 = 2\mathbf{x}^2 + \mathbf{a}^2 + \mathbf{b}^2 + 2\mathbf{ax} + 2\mathbf{bx},$$

and so $MX^2 + MZ^2 = MY^2 + MU^2$.

2. In the second part of the solution we defined the points A, B, C to the point P such that the segments PA, PB, PC are pairwise orthogonal. In fact, on the circle determined by A, B, C there are infinitely many triples A', B', C' having the same property.

If a cone has three pairwise perpendicular apothems, then it has infinitely many. Moreover, every apothem is the member of such a triple. (See [25].)

3. The planar version of the problem is the following: Let P be an inner point of a circle and let A, B be points on the circle such that PA and PB are perpendicular, then the vertices Q of the rectangles $PAQB$ lie on a circle.

1978/3. *The set of all positive integers is the union of two disjoint subsets:*
$$F = \{f(1), f(2), \ldots, f(n), \ldots\} \quad and$$
$$G = \{g(1), g(2), \ldots, g(n), \ldots\}$$
where
$$f(1) < f(2) < \ldots < f(n) < \ldots \quad and \quad g(1) < g(2) < \ldots < g(n) < \ldots,$$
(1) $\qquad g(n) = f(f(n)) + 1 \quad for\ every\ n \geq 1.$
Determine $f(240)$.

First solution. In principle, the values of f and g can be calculated, but it would be long and tedious. We shorten our way with some observations:

Let us count the number of integers between 1 and $g(n)$. The values $g(1)$, $g(2)$, ..., $g(n-1)$ precede $g(n)$ (this is $n-1$ integers), moreover as $g(n)-1 = f(f(n))$, the $f(n)$ values of f: $f(1)$, $f(2)$, ..., $f(f(n))$, thus we listed all numbers between 1 and $g(n)$, hence
(2) $\qquad g(n) = f(n) + n.$

Combining with (1) this gives
(3) $\qquad f(f(n)) = f(n) + n - 1.$

G cannot contain two consecutive numbers, because by (1) we have that $f(f(n))$ precedes $g(n)$. Thus the number following $g(n)$ is $f(f(n)+1)$:
(4) $\qquad f(f(n)+1) = f(f(n)) + 2.$

The formulas (1)–(4) make the evaluation of the functions easier. 1 has to be in F as (2) implies $g(1) = f(1) + 1$, thus $f(1) = 1$ and so $g(1) = 2$. By (4)
$$f(2) = f(f(1)+1) = f(f(1)) + 2 = 3.$$

The following values are calculated by (3):
$$f(3) = f(f(2)) = f(2) + 1 = 4,$$
$$f(4) = f(f(3)) = f(3) + 2 = 6,$$
$$f(6) = 9,\ f(9) = 14,\ f(14) = 22,\ f(22) = 35,\ f(35) = 56,\ f(56) = 90.$$

Now, applying (4) again:

$$f(57) = f(f(35) + 1) = f(f(35)) + 2 = 92.$$

And from (3):

$$f(92) = f(f(57)) = f(57) + 56 = 148,$$

$$f(148) = 239, f(239) = 386.$$

Finally, (4) gives

$$f(240) = f(f(148) + 1) = f(f(148)) + 2 = 388,$$

and we answered the question.

Second solution. Although this solution is much more complicated than the previous one, it gives a better view to the background of the problem.

First, we show that the conditions uniquely determine the functions f and g. It is enough to show that f is uniquely determined, because by (2) f determines g.

We prove by induction. By the previous proof $f(1) = 1$. Now, assume that the value of $f(n)$ is uniquely determined for every $n < k$ for some positive integer k. Let s denote the smallest positive integer that is not among the numbers $f(1)$, $f(2), \ldots, f(k-1), g(1), g(2), \ldots, g(k-1)$. $f(k) \geq s$, because all the numbers smaller than s are among the numbers listed above; $f(k) > s$ cannot hold, because by the monotonity of f it cannot belong to F, and by (2) we conclude that $g(k)$ is greater than s, so it cannot belong to G, either, contradicting the conditions of the problem. Thus $f(k) = s$, and we proved that f and g are uniquely determined.

So if we produce a pair of functions satisfying the conditions of the problem, that would be the unique pair of this kind.

We shall utilize the following theorem: if α and β are positive irrational numbers such that

(5)
$$\frac{1}{\alpha} + \frac{1}{\beta} = 1$$

holds, then the set of integers is the disjoint union of

(6) $\quad \{[\alpha], [2\alpha], \ldots, [n\alpha], \ldots\}$ and $\{[\beta], [2\beta], \ldots, [n\beta], \ldots\}$.

(See [26].)

We shall use the theorem when 1, α, β are three consecutive numbers of a geometric progression, i.e. $\beta = \alpha^2$, and by (5) in this case α is the positive root of the equation:

(7)
$$\alpha^2 - \alpha - 1 = 0.$$

$\alpha = \dfrac{1 + \sqrt{5}}{2} = 1{,}6180\ldots$, $\beta = \alpha^2 = \dfrac{3 + \sqrt{5}}{2} = 2{,}6180\ldots$ Since α and β are greater than 1 the sequences in (6) are strictly increasing.

So our choice for
$$f(n) = [n\alpha], \qquad g(n) = [n\alpha^2].$$
looks convenient.

To prove that our choice works, we have to show that (1) holds, i.e.

(8) $[n\alpha^2] = [[n\alpha]\alpha] + 1.$

$[n\alpha] < n\alpha$, so $[n\alpha]\alpha < n\alpha^2$, implying that
$$[[n\alpha]\alpha] \leq [n\alpha^2],$$
Equality cannot hold, because by (6) the l.h.s. and the r.h.s. belong to different sets, hence

(9) $[[n\alpha]\alpha] < [n\alpha^2].$

Reordering the inequality
$$n\alpha - [n\alpha] < 1 < \alpha,$$
$n\alpha < \alpha + [n\alpha]$ follows and as (7) implies $\alpha^2 - \alpha = 1$, we get
$$n\alpha < \alpha + (\alpha^2 - \alpha)[n\alpha].$$
Dividing this inequality by α, after reordering:
$$n + [n\alpha] < \alpha[n\alpha] + 1,$$
$$[n(1 + \alpha)] < \alpha[n\alpha] + 1$$
holds, or substituting $1 + \alpha = \alpha^2$:
$$[n\alpha^2] < [n\alpha]\alpha + 1,$$
hence
$$[n\alpha^2] \leq [[n\alpha]\alpha] + 1.$$
Considering (9) we have:
$$[[n\alpha]\alpha] < [n\alpha^2] \leq [[n\alpha]\alpha] + 1.$$
Since there is no integer between two consecutive integers, there is an equality at the second place. Hence (8) holds.

Our result implies
$$f(240) = [240\alpha] = [388,3281\ldots] = 388.$$

Remark. There are several ways to determine the values of the functions step by step. We mention the following:

There is a connection between the Fibonacci sequence defined by $a_1 = 1$, $a_2 = 1$, $a_{n+1} = a_n + a_{n-1}$ ($n > 1$) and the function $f(m)$:

(10) $f(a_n + 1) = a_{n+1} + 1.$ ($n > 1$)

Since $f(1) = 1$, $f(2) = 3$, $f(3) = 4$, (10) holds for the first few values. Using the recurrence formulas and (3):
$$a_{n+2} = f(a_{n+1} + 1) - 1 = f(f(a_{n+1})) - 1 =$$
$$= f(a_n + 1) - 1 + a_n = a_{n+1} + a_n,$$
so the connection is really true.

1978/4. *In the triangle ABC we have $AB = AC$. A circle is tangent internally to the circumcircle of the triangle and also to sides AB, AC at P, Q respectively. Prove that the midpoint of PQ is the centre of the incircle of the triangle.*

First solution. Let k denote the circle tangent to the circumcircle through P and Q. The point of tangency of k and the circumcircle is denoted by T. T is on the axis of symmetry of the triangle thus AT is a diameter of the circumcircle.

Let M denote the point of intersection of PQ and AT. We have to show that M is the centre of the incircle, as by the symmetry M is the midpoint of PQ (see *Figure 1978/4.1*).

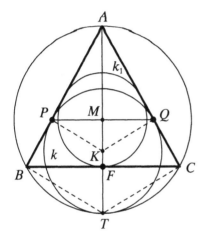

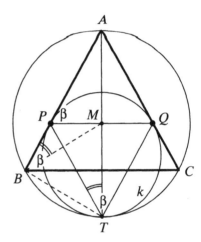

Figure 78/4.1   Figure 78/4.2

Thales' theorem implies that BT and CT are orthogonal to AB and AC, respectively. Let K denote the centre of the circle, and so PK and QK are orthogonal to the sides AB and AC, respectively. This implies that the deltoids $ABTC$ and $APKQ$ are homothetic with centre A, hence if F denotes the midpoint of BC, then

$$\frac{AT}{AF} = \frac{AK}{AM}.$$

Now, apply the enlargement with ratio $\dfrac{AT}{AF}$ and centre A to the circle k; this maps the circle k touching the lines AB and AC to a circle touching them. As the image of T is F at the enlargement, the image of the circle contains F hence the image is k_1, the incircle. Thus K is mapped to M, hence the centre of k_1 is M, and this is what we wanted to show.

Second solution. We use the notations of the first solution. In order to prove our statement it is enough to show that BM bisects $\angle ABC$, because in this case M is the point of intersection of two bisectors, hence the centre of the incircle (see *Figure 1978/4.2*). In the circle k we have $\angle PTQ = \angle APQ = \beta$ because they subtend the same arc. Moreover, $PBTM$ can be inscribed in a circle as its two opposite angles are right angles, hence $\angle PBM = \angle PTM$, because they subtend the same arc, and as $\angle PTM = \dfrac{\beta}{2}$, $\angle PBM$ equals $\dfrac{\beta}{2}$, as well; now, $\angle PBC = \beta$, hence $\angle MBC = \dfrac{\beta}{2}$. Thus BM is a bisector.

Third solution. We show that the statement holds for an arbitrary triangle, we do not need the condition $AB = AC$.

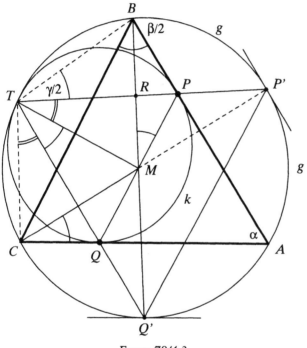

Fugre 78/4.3

Let k denote the circle touching AB, AC and g, the circumcircle of ABC, moreover denote by T the point of tangency of the two circles. T is the centre of enlargement of the two circles, hence Q' and P', the images of Q and P at the enlargement, are the midpoints of the arcs AC and AB, because the tangents AC and AB are parallel to the corresponding tangents (see *Figure 1978/4.3*).

As γ subtends the arc AB, $\dfrac{\gamma}{2}$ subtends its half, the arc BP', hence $\angle BTP' = \dfrac{\gamma}{2}$. Similarly, $\angle Q'TC = \dfrac{\beta}{2}$. The point of intersection of the bisector BQ' and the segment PQ is a single point M, as in the triangle $P'TQ'$ the side $P'Q'$ is parallel to PQ, and the points Q' and R (the point of intersection of the lines BQ' and TP'), are on different sides of the PQ line.

In the triangle PMB the exterior angle at P is $90° - \dfrac{\alpha}{2}$, hence $\angle PMB = 90° - \dfrac{\alpha}{2} - \dfrac{\beta}{2} = \dfrac{\gamma}{2}$, and so $\angle PTB$ and $\angle PMB$ are equal. $PMTB$ is a cyclic quadrangle, hence $\angle PTM = \angle PBM = \dfrac{\beta}{2}$. $\angle CTB$ is the external angle of α, thus

$$\angle CTB = \beta + \gamma = \angle BTP' + \angle P'TM + \angle MTQ' + \angle Q'TC = \frac{\gamma}{2} + \frac{\beta}{2} + \angle MTQ + \frac{\beta}{2},$$

and $\angle MTQ = \dfrac{\gamma}{2}$. This implies that $\angle MTC = \dfrac{\beta}{2} + \dfrac{\gamma}{2} = 90° - \dfrac{\alpha}{2} = \angle MQA$, thus

$MTCQ$ is an inscribed quadrangle, implying $\angle MTQ = \angle MCQ = \dfrac{\gamma}{2}$. Hence MC is the bisector of γ, M is the point of intersection of the two bisectors, and so the centre of the inscribed circle of ABC. Thus M lies on the bisector of the isosceles triangle PAQ at A, and so M bisects the segment PQ.

Fourth solution. Using well known trigonometric formulas we can give a simple proof to the generalized version of the problem.

Let T, P, and Q denote the points of tangency of the circle k and the circumcircle, and the segments AB and AC respectively, O and R the centre and the radius of the circumcircle, r the radius of the incircle, and S and ϱ the centre and the radius of the circle k (see *Figure 1978/4.4*). $\angle QAO = 90° - \beta$ hence $|\angle OAS| = \omega = \left| \dfrac{\alpha}{2} - (90° - \beta) \right| =$

$= \left| \dfrac{\beta - \gamma}{2} \right|$. (Here, $\beta > \gamma$.)

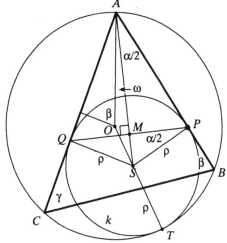

Figure 78/4.4

Since O, S, T are collinear, $OS = = R - \varrho$; the right triangle AQS satisfies $AS = \dfrac{\varrho}{\sin \frac{\alpha}{2}}$. Applying the Law of Cosine for the triangle AOS we get:

$$OS^2 = OA^2 + AS^2 - 2OA \cdot AS \cos \omega,$$

$$(R - \varrho)^2 = R^2 + \frac{\varrho^2}{\sin^2 \frac{\alpha}{2}} - \frac{2R\varrho}{\sin \frac{\alpha}{2}} \cos \frac{\beta - \gamma}{2}.$$

After rearranging we obtain:

$$\varrho \cos^2 \frac{\alpha}{2} = 2R \sin \frac{\alpha}{2} \left(\cos \frac{\beta - \gamma}{2} - \sin \frac{\alpha}{2} \right) =$$

$$= 2R \sin \frac{\alpha}{2} \left(\cos \frac{\beta - \gamma}{2} - \cos \frac{\beta + \gamma}{2} \right) = 2R \sin \frac{\alpha}{2} \sin \frac{\beta}{2} \sin \frac{\gamma}{2}.$$

Using the formula $2R \sin \dfrac{\alpha}{2} \sin \dfrac{\beta}{2} \sin \dfrac{\gamma}{2} = r$:

(1)
$$\varrho = \frac{r}{\cos^2 \frac{\alpha}{2}}.$$

Let M denote the point of intersection of the segments AS and PQ (this is the midpoint of PQ). In the right triangle PMS the angle $\angle SPM = \dfrac{\alpha}{2}$ implying $MS = \varrho \sin \dfrac{\alpha}{2}$, $MP = \varrho \cos \dfrac{\alpha}{2}$ The geometric mean theorem applied to the right triangle SPA gives $MP^2 = AM \cdot MS$, and using (1) we get:

$$AM = \frac{MP^2}{MS} = \frac{\varrho^2 \cos^2 \frac{\alpha}{2}}{\varrho \sin \frac{\alpha}{2}} = \frac{r}{\sin \frac{\alpha}{2}}.$$

As the distance of the vertex A from the centre of the incircle is $\dfrac{r}{\sin \frac{\alpha}{2}}$, M is the centre of the incircle and this is what we wanted to prove.

Remarks. 1. The third solution is essentially the same as the second solution of the problem 1969/4. If there, we identify the points D and B we get this third solution.

2. A careful investigation of *Figure 1978/4.3* suggests the obvious solution of the problem: we have to show that the points P, Q and M are collinear. This is the straightforward consequence of Pascal's theorem:

Let 1, 2, 3, 4, 5, 6 be six arbitrary points of a conic (e.g. a circle), then the points of intersection of the lines 12 and 45, 23 and 56 moreover 34 and 61 are collinear.

On *Figure 1978/4.3* let $A = 1$, $B = 2$, $Q' = 3$, $T = 4$, $P' = 5$, $C = 6$. By Pascal's theorem P, the point of intersection of AB and TP', M, the point of intersection of BQ' and $P'C$ and Q, the point of intersection of $Q'T$ and CA are collinear.

1978/5. *Let $\{a_k\}$ be a sequence of distinct positive integers ($k = 1, 2, \ldots,$ $n, \ldots$). Prove that for every positive integer n*

(1)
$$\sum_{k=1}^{n} \frac{a_k}{k^2} \geq \sum_{k=1}^{n} \frac{1}{k}.$$

Solution. We start with a lemma: let $x_1, x_2, \ldots, x_n$; $y_1, y_2, \ldots, y_n$ be real numbers, where $x_1 \geq x_2 \geq \ldots \geq x_n$, $y_1 \leq y_2 \leq \ldots \leq y_n$, moreover $z_1, z_2, \ldots, z_n$ a permutation of the y_i-s, then

(2)
$$\sum_{i=1}^{n} x_i y_i \leq \sum_{i=1}^{k} x_i z_i.$$

In (2) equality holds if the order of the z_i-s agrees with the order of the y_i-s. If to the contrary, we assume that they first differ at the k-th place, i.e. $y_1 = z_1$,

$y_2 = z_2, \ldots, y_{k-1} = z_{k-1}$, but $y_k \neq z_k$. Let $z_k = y_r$ and $y_k = z_s$, where r and s are greater than k, $r > k$ and $s > k$. Using this notation

$$x_k z_k + x_s z_s = x_k y_r + x_s y_k.$$

Now, change the left hand side by substituting $x_k z_k + x_s z_s = x_k y_r + x_s y_k$ by $x_k y_k + x_s y_r$. This way we increased (did not decrease) the sum on the left, because

$$(x_k y_k + x_s y_r) - (x_k y_r + x_s y_k) = (x_k - x_s)(y_k - y_r) \geq 0.$$

This implies that the sum $\sum x_i y_i$, where the x_i-s are in decreasing order is the smallest if the y_i-s are in increasing order. There is a minimal among the above sums as there are finitely many of them and that can only occur when the y_i-s are in increasing order, otherwise we could decrease the sum using our arguments.

We utilize the lemma in the following way: for a given n let

$$a_{i_1} \leq a_{i_2} \leq \ldots \leq a_{i_n}$$

be the first k elements in increasing order. But

$$\frac{1}{1^2} > \frac{1}{2^2} > \frac{1}{3^2} > \ldots > \frac{1}{n^2},$$

and hence by (2)

(3) $$\sum_{k=1}^{n} \frac{a_k}{k^2} \geq \sum_{k=1}^{n} \frac{a_{i_k}}{k^2}.$$

$a_{i_1} \geq 1$, $a_{i_2} \geq 2$, ..., $a_{i_k} \geq k$, ..., $a_{i_n} \geq n$, and so (3) implies that

$$\sum_{k=1}^{n} \frac{a_k}{k^2} \geq \sum_{k=1}^{n} \frac{k}{k^2} = \sum_{k=1}^{n} \frac{1}{k},$$

and this is what we wanted to prove.

Remarks. The full version of the lemma that we proved in the solution of problem 1975/1 is the following (see also [35]):

If $x_1, x_2, \ldots, x_n$ and $z_1, z_2, \ldots, z_n$ are real n-tuples, then the sum

$$S = \sum_{i=1}^{n} x_i z_i$$

is maximal if the tuples are ordered on the same way and minimal if they are in reversed order.

2. A consequence of the statement of the problem is that the series $\sum_{k=1}^{\infty} \frac{a_k}{k^2}$

is divergent (goes to infinity), because the same holds for the series $\sum_{k=1}^{\infty} \frac{1}{k}$. This is not true if there are only finitely many distinct a_i-s. If a is the largest one,

$$\sum_{k=1}^{n} \frac{a_k}{k^2} \le a \left(\frac{1}{1^2} + \frac{1}{2^2} + \ldots + \frac{1}{n^2} \right) < a \left(1 + \frac{1}{1 \cdot 2} + \frac{1}{2 \cdot 3} + \ldots + \frac{1}{n(n-1)} \right) =$$

$$= a \left(1 + \frac{1}{1} - \frac{1}{2} + \frac{1}{2} - \frac{1}{3} + \ldots + \frac{1}{n-1} - \frac{1}{n} \right) = a \left(2 - \frac{1}{n} \right) < 2a,$$

hence every partial sum of the series $\sum_{k=1}^{\infty} \frac{a_k}{k^2}$ is smaller than $2a$, thus the series is convergent and its sum is smaller than $2a$. If every a_k is 1, then we get the well known sum: $\sum_{1}^{\infty} \frac{1}{k^2} = \frac{\pi^2}{6} = 1,6449 \ldots$.

1978/6. *An international society has its members from six different countries. The list of members has 1978 names, numbered 1, 2, ..., 1978. Prove that there is at least one member whose number is the sum of the numbers of two members from his own country, or twice the number of a member from his own country.*

First solution. The statement of the problem is equivalent to the following: there are two scientists from the same country such that the difference of their number belongs to a scientist of their country.

Assume to the contrary that the statement is false. Let A, B, C, D, E and F denote the countries. There is a country with at least 330 scientists, e.g. A, because $6 \cdot 329 = 1974 < 1978$. Their numbers are:

$$a_1 < a_2 < \ldots < a_{330}.$$

By our assumption no scientist with number

(1) $$a_2 - a_1, a_3 - a_1, \ldots, a_{330} - a_1$$

is from A, instead they are distributed among the other 5 countries.

As $5 \cdot 65 = 325 < 329$, there are at least 66 scientists from the same country different from A, e.g. from B, from the list (1). Let their numbers be

$$b_1 < b_2 < \ldots < b_{66}.$$

Now, none of the scientists with number

(2) $$b_2 - b_1, \ b_3 - b_1, \ \ldots, \ b_{66} - b_1$$

can be from B, but they are not from A either, because e.g. $b_2 - b_1 = (a_k - a_1) - (a_j - a_1) = a_k - a_j$ were in A, contradicting our assumptions.

Now, since $4 \cdot 16 = 64 < 65$, there are at least 17 scientists from the same country different from A, and B e.g. from C from the list (2) with numbers

$$c_1 < c_2 < \ldots < c_{17}.$$

None of the scientist with number

(3) $$c_2 - c_1, c_3 - c_1, \ldots, c_{17} - c_1$$

can be from C, but not from B as e.g. $c_k - c_1 = (b_s - b_1) - (b_r - b_1) = b_s - b_r$ would hold contradicting our assumptions about B. Similarly, it cannot belong to A, either.

As $3 \cdot 5 = 15 < 16$, there are at least 6 scientists from the list in (3) belonging to the same country, let us say from D. They are

$$d_1 < d_2 < \ldots < d_6.$$

Using our previous arguments, it is clear that none of the five scientists

(4) $$d_2 - d_1, d_3 - d_1, \ldots, d_6 - d_1$$

belong to A, B, C or D. So there are at least 3 scientists from E, because $2 \cdot 2 = 4 < 6$:

$$e_1 < e_2 < e_3.$$

The scientists $e_2 - e_1$ and $e_3 - e_1$ are not from the countries E, A, B, C, D, hence they are from F:

$$f_1 = e_2 - e_1 \quad \text{and} \quad f_2 = e_3 - e_1.$$

But now, the scientist with number $f_1 - f_2$ cannot be from any of the countries, providing a contradiction, hence the statement is true.

Second solution. We prove a graph theoretical generalisation of the problem.

Let

(5) $$n_k = k! \left(1 + \frac{1}{1!} + \frac{1}{2!} + \ldots + \frac{1}{k!} \right) + 1.$$

If we colour the edges of the complete graph with n_k vertices by k colours, then there are three edges in the graph that form a triangle. Of course, we do not have to use all k colours...

We prove our statement by induction on k. Suppose first that $k = 1$, $n_1 = 3$. Here, the existence of the monochromatic triangle is obvious. Now, assume that the statement holds for some k and we have to prove that it is true for $k + 1$.

We shall use the following consequence of (5):

(6) $$(k+1)(n_k - 1) = n_{k+1} - 2 < n_{k+1} - 1.$$

Choose an arbitrary vertex P form the graph with n_{k+1} vertices and coloured by $k + 1$ colours. P is the endpoint of $n_{k+1} - 1$ edges, coloured by the $k + 1$ colours. There must be n_k many edges of the same colour, as by (6)

$(n_k - 1)(k + 1)$ is smaller then the edges starting from P. Let these edges be coloured red. The other vertices form a complete graph G with n_k vertices. If G contains a red edge, then we have a red triangle.

If G does not contain a red edge, then it is a complete graph with n_k vertices coloured by k colours, thus by the assumptions G contains a monochromatic triangle. Thus we proved our theorem.

Now, we can solve the original problem: Let us assign a vertex of a complete graph to every participant, and label it by the number of the scientists. Assign a colour to all six countries. We colour the edge between i and j by the colour of the country where the scientist with number $|i - j|$ belongs to. Since $n_6 = 1958 < 1978$, the graph contains a monochromatic triangle with vertices $i < $ $< j < k$. In this case the scientists with number $j - i$, $k - j$, $k - i$ are from the same country, and

$$(j - i) + (k - j) = k - i,$$

as we wanted. If $j - i = k - j$, then $k - i$ is twice as $j - i$, so it fits the conditions of the problem.

Remarks. 1. The graph theoretic theorem is a Ramsey-type theorem ([13]).

2. In the literature, the r.h.s. of (5) is denoted by $[ek!] + 1$, where e is the base of the natural logarithm. The notation is relevant, as:

(7) $$ek! + 1 - n_k = ek! - (n_k - 1) < 1.$$

his can be easily verified, since

$$e = 1 + \frac{1}{1!} + \frac{1}{2!} + \frac{1}{3!} + \ldots,$$

$$ek! = k! \left(1 + \frac{1}{1!} + \frac{1}{2!} + \ldots + \frac{1}{k!}\right) + \frac{1}{k+1} + \frac{1}{(k+1)(k+2)} + \ldots$$

$$= n_k - 1 + \frac{1}{k+1} + \frac{1}{(k+1)(k+2)} + \ldots$$

If $k > 1$,

$$ek! - (n_k - 1) = \frac{1}{k+1} + \frac{1}{(k+1)(k+2)} + \ldots < \frac{1}{k+1} + \frac{1}{(k+1)^2} + \ldots =$$

$$= \frac{1}{k+1} \cdot \frac{1}{1 - \frac{1}{k+1}} = \frac{1}{k} < 1,$$

and as (7) is obvious for $k = 1$, (7) holds for every n.

1979.

1979/1. *Let p and q be positive integers such that*

$$\frac{p}{q} = 1 - \frac{1}{2} + \frac{1}{3} - \frac{1}{4} + \ldots - \frac{1}{1318} + \frac{1}{1319}.$$

Prove that 1979 divides p.

Solution. Apply the following transformation to the sum:

$$\frac{p}{q} = 1 - \frac{1}{2} + \frac{1}{3} - \frac{1}{4} + \ldots - \frac{1}{1318} + \frac{1}{1319} = 1 + \frac{1}{2} + \frac{1}{3} + \ldots + \frac{1}{1319} -$$

$$- 2\left(\frac{1}{2} + \frac{1}{4} + \frac{1}{6} + \ldots + \frac{1}{1318}\right) = 1 + \frac{1}{2} + \frac{1}{3} + \ldots + \frac{1}{1319} -$$

$$- \left(1 + \frac{1}{2} + \frac{1}{3} + \ldots + \frac{1}{659}\right) = \frac{1}{660} + \frac{1}{661} + \ldots + \frac{1}{1319} =$$

$$= \left(\frac{1}{660} + \frac{1}{1319}\right) + \left(\frac{1}{661} + \frac{1}{1318}\right) + \ldots + \left(\frac{1}{989} + \frac{1}{990}\right) =$$

$$= 1979 \left(\frac{1}{660 \cdot 1319} + \frac{1}{661 \cdot 1318} + \ldots + \frac{1}{989 \cdot 990}\right).$$

The sum in the parentheses is $\frac{a}{b}$, where a is a positive integer and $b = 660 \cdot 661 \cdot$
$\cdot \ldots \cdot 1319$. As 1979 is a prime and every factor of b is smaller than 1979, b
and 1979 are coprime. Thus $\frac{p}{q} = \frac{1979a}{b}$ implies that $pb = 1979aq$ and so 1979
divides p.

Remark. The problem can be generalized for primes of the form $3k + 2$: if

$$\frac{p}{q} = 1 - \frac{1}{2} + \frac{1}{3} - \ldots - \frac{1}{2k} + \frac{1}{2k+1},$$

then $3k + 2$ divides p.

1979/2. *A prism with pentagons* $A_1A_2A_3A_4A_5$, *and* $B_1B_2B_3B_4B_5$ *as
top and bottom faces is given. Each side of the two pentagons and each of the
25 segments* A_iB_j, *(i, j = 1, 2, ..., 5) is coloured red or green. Every triangle
whose vertices are vertices of the prism and whose sides have all been coloured
has two sides of a different colour. Prove that all 10 edges of the top and bottom
faces have the same colour.*

Solution. We call the coloured segments in the problem edges. First we
show that all five edges of the top and bottom faces have the same colour. Suppose to the contrary that the bottom face has a red and a green edge.

Let A_2 be the common vertex of the red edge A_1A_2, and the green edge
A_2A_3 (see *Figure 1979/2.1*). There are three edges of the same colour (e.g.
red) among the edges between A_2 and the 5 vertices of the top face. As any
3 edges of a pentagon contain two adjacent vertices, there are two neighbour
vertices of the top face, B_i and B_k that are connected by a red edge with A_2. As
every triangle has two distinct colours, the edges A_1B_k and A_1B_i are green, for
otherwise we would have $A_1A_2B_i$ or $A_1A_2B_k$ as red triangles. But then in the

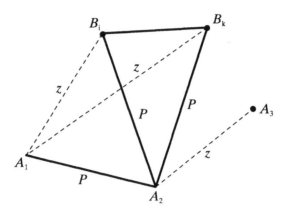

Figure 79/2.1

triangle $B_iB_kA_1$ the edge B_iB_k has to be red, a contradiction, because then the triangle $A_2B_iB_k$ is red. So the edges of the bottom face are of the same colour. By symmetry the same holds for the top face.

Now, it only remained to show that the top and bottom faces have the same colour. Let the bottom face be red. One of the end points of the red edge A_1A_2 is connected to the B_i vertex with green. So one of them (e.g. A_1) is the endpoint of 3 green edges. Two of these green edges end in adjacent vertices, so the edge connecting them (on the top face) is red. But then, all edges of the top face are red as we proved above.

Remark. We only used the fact that the top and bottom faces have odd many vertices, we have not even used that they are of the same size. In fact we proved the following graph theoretical statement:

Let p and q be two odd cycles in the graph G such that there is no edges between the non-adjacent edges of p and q, but every vertex of q is connected with every vertex of p. Colour the vertices of G by two colours such that there is no monochromatic triangle. Then the edges of p and q have the same colour.

1979/3. *Let k_1 and k_2 be two circles on the plane and let A denote one of their points of intersection. Starting simultaneously from A, two points, P_1 and P_2 move with constant speed, each traveling along its own circle in the same sense. The two points return to A simultaneously after one revolution. Prove that there is a fixed point P in the plane such that, at any time, the distance from P to the moving points are equal.*

First solution. Let O_1 and O_2 denote the centres of k_1, and k_2, r_1 and r_2 their radii, respectively. After examining some special cases (see *Figure 1979/3.1*), our guess is that P is the reflected image of A to the perpendicular bisector of the O_1O_2 segment.

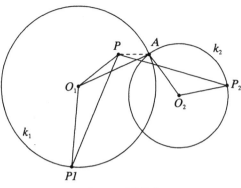

Let P_1 and P_2 be a location of the moving points. Then $\angle AO_1P_1 = \angle AO_2P_2$. The triangle AO_1P is the symmetric image of the triangle PO_2A, $\angle PO_1A = \angle PO_2A$, hence $\angle PO_1P_1 = \angle PO_2P_2$. Moreover, the symmetry implies that $AO_1 = PO_2 = r_1$ and $AO_2 = PO_1 = r_2$, too. It follows that two sides and their angle agree in the triangles PO_1P_1 and PO_2P_2, hence they are congruent and so $PP_1 = PP_2$.

Figure 79/3.1

Note that the length of the two segments equal even if the points P, O_1, P_1, and the points P, O_2, P_2 are collinear. The length of the segments in this case are $r_1 + r_2$ or $|r_1 - r_2|$.

Second solution. We start with the following observation: if the similar triangles OA_1B_1, OA_2B_2, OA_3B_3 are of the same orientation and the points A_1, A_2, A_3 are collinear, then the points B_1, B_2, B_3 are collinear, as well. (See *Figure 1979/3.2*). This is simply true, because the rotation stretching with centre O and ratio OB_1/OA_1 and angle A_1OB_1 map the points A_1, A_2, A_3 to B_1,

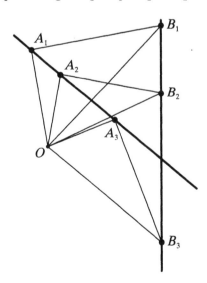

Figure 79/3.2

B_2, B_3, respectively. Since the rotation stretching maps line to line, the line containing A_1, A_2, A_3 is mapped to the line through B_1, B_2 and B_3. This holds even if the triples O, A_1, B_1; O, A_2, B_2; O, A_3, B_3 are collinear.

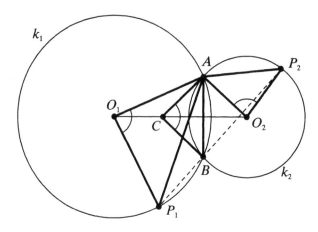

Figure 79/3.3

Now, let us assume that the points P_1 and P_2 both travelled an arbitrary angle α, $(0 < \alpha < 2\pi)$ around the centres of the circles (see *Figure 1979/3.3*). Thus the triangles AO_1P_1, AO_2P_2 (possibly degenerated) are similar isosceles triangles of the same orientation. As the perpendicular bisector of the segment AB is the line O_1O_2 (here, B denotes the other point of intersection of the two circles), there is a point C on O_1O_2 such that the triangle ACB is similar to the triangle AO_1P_1 and they are oriented in the same way.

As the triangles AO_1P_1, AO_2P_2 and ACB are similar triangles of the same orientation, and the points O_1, O_2, C are collinear, our introductory remark implies that P_1, P_2 and B are always collinear. So we only have to show that at any stage the perpendicular bisector of P_1P_2 contains a fix point. The second point of intersection of the perpendicular to AB at A with the circle k_1 is X and with the circle k_2 is Y (see *Figure 1979/3.4*). Let P denote the midpoint of XY, this is independent of the points.

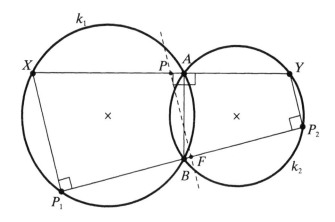

Figure 79/3.4

The theorem of Thales implies that BX is a diameter in k_1, and BY is a diameter in k_2, hence the quadrangle XP_1P_2Y is a right trapezium. The perpendicular bisector of P_1P_2 is the median of the trapezium, hence it contains the midpoint of XY.

Remark. The basic observation of the second solution is that the points P_1, P_2 and B are collinear. We can verify this by calculating the angles, but using the rotation stretching needs less discussion. Of course, analytic or trigonometric method may also lead to the solution.

1979/4. *Given a plane π, a point P in the plane and a point Q not in the plane. Find all points R of the plane π such that the ratio*

(1)
$$\frac{QP+PR}{QR}$$

is maximal.

Solution. Let Q' denote the orthogonal projection of Q to the plane π, and m denote the PQ' line; if $P=Q'$, then let m be an arbitrary line of the plane through Q'. We show that if R is not on m, then m has a point R' such that

$$\frac{QP+PR}{QR} < \frac{QP+PR'}{QR'}.$$

So for every point R not on m there is a point R' on m that gives a greater value at (1).

Indeed, if m does not contain R let R' denote the point on m such that $Q'R'=Q'R$ and R' is not on the same ray as P (see *Figure 1979/4.1*). As the triangles $QQ'R$ and $QQ'R'$ are congruent,

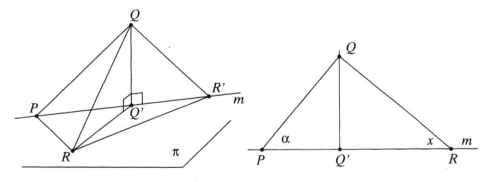

Figure 79/4.1   Figure 79/4.2

$QR=QR'$, and from the triangle inequality $PR<PQ'+Q'R=PQ'+ +Q'R'=PR'$. Hence

$$\frac{QP+PR}{QR} = \frac{QP+PR}{QR'} < \frac{QP+PR'}{QR'}.$$

Thus the points providing the maximum are on m.

From now, we assume that m contains R (see *Figure 1979/4.2*). Let $\angle QPR = \alpha$ and $\angle QRP = x$, and apply the Law of Sines for the triangle QPR:

$$\frac{QP+PR}{QR} = \frac{QP}{QR} + \frac{PR}{QR} = \frac{\sin x}{\sin \alpha} + \frac{\sin(\alpha+x)}{\sin \alpha} =$$

$$= \frac{1}{\sin \alpha} 2 \sin \frac{2x+\alpha}{2} \cos \frac{\alpha}{2} = \frac{1}{\sin \frac{\alpha}{2}} \sin \left(x + \frac{\alpha}{2} \right).$$

Our expression is maximal if $\sin \left(x + \frac{\alpha}{2} \right) = 1$, that is if $x + \frac{\alpha}{2} = 90°$, $x = 90° - \frac{\alpha}{2}$. Hence $\angle PQR = 180° - (x+\alpha) = 90° - \frac{\alpha}{2} = x$, and the triangle PQR is an isosceles triangle, $QP = RP$.

Note that there are two points on m of distance QP from P. The denominator of (1) is the same in both cases, but if $P \neq Q'$, QR is smaller (i.e. (1) is greater), where the angle QPR is acute. The unique solution, R is the endpoint of the segment PR of length PQ on the PQ' line on the ray not containing Q'.

If $P = Q'$, our arguments show that the points R are the points of the circle in π with centre P and radius PQ.

1979/5. *Find all real numbers b for which there exist nonnegative real numbers x_1, x_2, x_3, x_4, x_5 satisfying*

(1)
$$\sum_{k=1}^{5} k x_k = b, \quad \sum_{k=1}^{5} k^3 x_k = b^2, \quad \sum_{k=1}^{5} k^5 x_k = b^3.$$

First solution. By (1)

$$b^2 \sum_{k=1}^{5} k x_k - 2b \sum_{k=1}^{5} k^3 x_k + \sum_{k=1}^{5} k^5 x_k = b^3 - 2b^3 + b^3 = 0,$$

that is

$$0 = \sum_{k=1}^{5} x_k (b^2 k - 2bk^3 + k^5) = \sum_{k=1}^{5} k(b - k^2)^2 x_k.$$

This latter equation is equivalent to

(2) $(b-1)^2 x_1 + 2(b-4)^2 x_2 + 3(b-9)^2 x_3 + 4(b-16)^2 x_4 + 5(b-25)^2 x_5 = 0.$

The 5 terms on the l.h.s. are all non negative. This can only hold if

$$x_1 = x_2 = x_3 = x_4 = x_5 = 0, \qquad b = 0,$$

or if $x_k \neq 0$, then in the k-th term $b - k^2 = 0$, i.e. $b = k^2$. In this case in the other terms $b - k^2 \neq 0$ holds, and so $x_k = 0$. The table of the solutions:

x_1	x_2	x_3	x_4	x_5	b
0	0	0	0	0	0
1	0	0	0	0	1
0	2	0	0	0	4
0	0	3	0	0	9
0	0	0	4	0	16
0	0	0	0	5	25

Easy to verify that these values clearly satisfy (1).

Second solution. The Cauchy-inequality for arbitrary real numbers, $a_1, a_2, \ldots, a_n; b_1, b_2, \ldots, b_n$ states that

(3)
$$\sum_1^n a_k^2 \sum_1^n b_k^2 \geq \left(\sum_1^n a_k b_k \right)^2 ,$$

where equality holds if and only if there is a t such that $a_k = t b_k$ ($k = 1, 2, \ldots, n$) or all $b_k = 0$ ([22]).

Set

$$a_k = \sqrt{k x_k}, \qquad b_k = \sqrt{k^5 x_k}.$$

Now (3) reads as:

$$\sum_1^5 k x_k \cdot \sum_1^5 k^5 x_k \geq \left(\sum_1^5 k^3 x_k \right)^2 ,$$

by (1) we have $b^3 \geq b^4$, thus equality holds in (3), hence there is a y such that

(4) $\sqrt{k x_k} = t\sqrt{k^5 x_k}$, $x_k(t^2 k^4 - 1) = 0$ ($k = 1, 2, \ldots, 5$)

or every $b_k = 0$, so every $x_k = 0$, and necessarily $b = 0$.

If in (4) $x_i \neq 0$ for some i, then $t = \dfrac{1}{i^2}$, but then by (4) for every other k, $x_k = 0$ ($k \neq i$). Substituting to (1) we obtain

$$i x_i = b, \qquad i^3 x_i = b^2, \qquad i^5 x_i = b^3,$$

implying $b = i^2$ and $x_i = i$; Thus the possible values of b are: 0, 1, 4, 9, 16 and 25; and these values satisfy the equation.

1979/6. *Let A and E be opposite vertices of an octagon. A frog starts at vertex A. From any vertex except E it jumps to one of the two adjacent vertices. When it reaches E it stops. Let a_n be the number of distinct paths of exactly n jumps ending at E. Prove that*

$$a_{2n-1} = 0, \qquad a_{2n} = \frac{1}{\sqrt{2}} \left(x^{n-1} - y^{n-1} \right) \qquad (n = 1, 2, 3, \ldots)$$

where $x = 2 + \sqrt{2}$ and $y = 2 - \sqrt{2}$.

Remark. A path of exactly n jumps is a sequence of the vertices P_0, P_1, ..., P_n that satisfy the following conditions:

I. $P_0 = A$, $P_n = E$;

II. for every i, $0 \leq i \leq n - 1$ P_i is distinct from E.

III. for every i, $0 \leq i \leq n - 1$, P_i and P_{i+1} are adjacent vertices.

Solution. The conditions in the remark are the complicated formulation of the following: the frog, starting from A reaches E with the n-th jump, but not sooner, and it can jump only to adjacent vertices.

Let A, B, C, D, E, F, G, H denote the vertices of the octagon (see *Figure 1979/6.1*). The vertices B and H can be reached from A only with odd many jumps, C and G with even many jumps, and so D and F by odd many jumps, finally E by even many jumps. Thus $a_n = 0$, if n is odd.

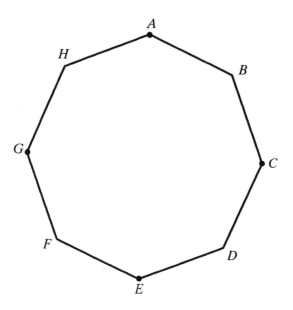

Figure 79/6.1

From now we assume that n is even. We try to express a_n with the number of paths shorter than n. Obviously $a_2 = 0$ and $a_4 = 2$, because the frog needs at least four jumps from A to E and it can do it in two ways. From now we assume that $n > 4$.

Let b_n denote the number of the paths of n jumps from C to E. By symmetry, there are b_n many paths from G to E.

After two jumps there are two cases. First case: the frog is in A. Second case: the frog is in C or G. In the first case there are $2a_{n-2}$ paths from A to E

as there are two ways to get to A from A with two jumps. In the second case there are b_{n-2} paths to E starting from C or G. Combining the cases gives:

(1) $$a_n = 2a_{n-2} + 2b_{n-2}.$$

Let us find a similar formula for b_n. Starting from C after two steps the frog (on a continuable path) is in C or in A. From A to E there are a_{n-2} many, from C to E there are b_{n-2} many $n-2$ long paths, but there are two ways to get from C to C, hence

(2) $$b_n = a_{n-2} + 2b_{n-2}.$$

Combining (1) and (2), after reordering we get

$$b_n = a_n - a_{n-2},$$

implying $b_{n-2} = a_{n-2} - a_{n-4}$. Substituting to (1) we get a recurrence formula for a_n:

(3) $$a_n = 4a_{n-2} - 2a_{n-4}.$$

Finally, we need to show that

(4) $$a_{2n} = \frac{1}{\sqrt{2}} \left(x^{n-1} - y^{n-1} \right)$$

satisfy (3) for every positive integer n. We do it by induction. For $n = 1$, $a_2 = 0$, for $n = 2$ $a_4 = 2$ that coincides with our results. Now, assume that (4) holds for $n - 1$ and we have to show that it is true for n, as well.

Observe that $x + y = 4$ and $xy = 2$, thus x and y are the roots of the equation

(5) $$z^2 - 4z + 2 = 0,$$

and this implies that x and y satisfy the equation:

$$z^{n-1} = 4z^{n-2} - 2z^{n-3}.$$

Using this:

$$a_{2n} = 4a_{2(n-1)} - 2a_{2(n-2)} = \frac{4}{\sqrt{2}} \left(x^{n-2} - y^{n-2} \right) - \frac{2}{\sqrt{2}} \left(x^{n-3} - y^{n-3} \right) =$$

$$= \frac{1}{\sqrt{2}} \left(\left(4x^{n-2} - 2x^{n-3} \right) - \left(4y^{n-2} - 2y^{n-3} \right) \right) = \frac{1}{\sqrt{2}} \left(x^{n-1} - y^{n-1} \right),$$

and this is what we wanted to prove.

Remark. Before the competition the jury considered a difficult version of the problem. Instead the "Prove that ..." part they thought about the "Determine the value of a_n", that is, the problem would be to find the value of (4). From (3) it is easy to produce (4) (see [27]); some students first observed from (4) that (3) holds and then they started to solve the problem.

1980.

In 1980 there was no International Mathematical Olympics. The Organisation of the Teachers of Mathematics, Physics and Chemistry of Finland (MAOL) invited the teams of England, Hungary and Sweden to compare their abilities. Here we present the problems of that competition.

1980/1. *Let* α, β, *and* γ *denote the angles of the triangle* ABC. *The perpendicular bisector of* AB *intersects* BC *at the point* X, *the perpendicular bisector of* AC *intersects it at* Y. *Prove that* $\tan \beta \cdot \tan \gamma = 3$ *implies* $BC = XY$.

Show that this condition is not necessary for $BC = XY$, *and give a sufficient and necessary condition.*

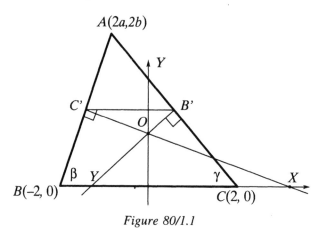

Figure 80/1.1

First solution. The most convenient way of solving the problem is the analytic method, as we can avoid to distinguish the several cases.

First, observe that for right triangles the problem does not make sense, because the perpendicular bisectors do not intersect the BC line or they intersect it in the same point. From now, we assume that ABC is not a right triangle.

Insert the ABC triangle into the coordinate system such that the coordinates of the vertices are: $A(2a, 2b)$, $B(-2, 0)$, $C(2, 0)$ (see *Figure 1980/1.1*); we may assume that $a \leq 0$. The coordinates of the midpoints of the sides AB and AC are: $C'(a - 1, b)$, $B'(a + 1, b)$; the side vectors: $(a + 1, b)$, $(a - 1, b)$, hence the equations of the perpendicular bisectors are:

$$(a+1)x + by = a^2 + b^2 - 1, \quad \text{and} \quad (a-1)x + by = a^2 + b^2 - 1.$$

They intersect the x axis (BC line) at the points

$$X\left(\frac{a^2 + b^2 - 1}{a + 1}, 0\right), \qquad Y\left(\frac{a^2 + b^2 - 1}{a - 1}, 0\right).$$

($a - 1 \neq 0$, since $a \leq 0$ and $a + 1 = 0$ means that the coordinates of A are $(-2, 2b)$ and then the triangle ABC is a right triangle). Thus, the XY distance is:

(1)
$$XY = \left| \frac{a^2 + b^2 - 1}{a - 1} - \frac{a^2 + b^2 - 1}{a + 1} \right| = \left| -\frac{2b^2}{a^2 - 1} - 2 \right|.$$

And as $BC = 4$, the sufficient and necessary condition required by the problem is

(2) $$\left| -\frac{2b^2}{a^2 - 1} - 2 \right| = 4, \quad \text{that is} \quad \left| -\frac{b^2}{a^2 - 1} - 1 \right| = 2.$$

Using our notations

$$\tan \beta = \frac{b}{a+1}, \qquad \tan \gamma = -\tan(180° - \gamma) = \frac{-b}{a-1},$$

hence $\tan \beta \cdot \tan \gamma = -\dfrac{b^2}{a^2 - 1}$, and the condition in (2) rewrites as

(3) $$|\tan \beta \tan \gamma - 1| = 2.$$

This splits into two parts, depending on the sign of the expression in the absolute value:

(4) $$\tan \beta \tan \gamma = 3,$$

and

(5) $$\tan \beta \tan \gamma = -1.$$

Thus both (4) and (5) are sufficient conditions, and together they are necessary and sufficient for $XY = BC$.

Second solution. Let us assume that ABC is an acute triangle. Use the notations of *Figure 1980/1.1*. The perpendicular bisectors at B' and C' intersect in O, the centre of the circumcircle, which is inside the triangle. The triangles $OB'C'$ and OXY are similar and O separates the points B' and Y, and C' and X. If $XY = BC$, as $B'C' = \dfrac{BC}{2}$, the ratio of the similarity of the two triangles is $2 : 1$. Hence — as the sum of the lengths of the altitudes of the triangles through O is $\dfrac{h_a}{3}$, the distance of O from BC is $\dfrac{h_a}{3}$. This is the same as the distance of the point of gravity of ABC from BC, hence the line through O and the point of gravity (that is the the Euler line of the triangle) is parallel to BC.

It is a well known theorem that the sufficient and necessary condition for the Euler line and BC being parallel is (see the remark): in case $\beta \neq \gamma$:

(6) $$\tan \beta \tan \gamma = 3$$

In case an acute triangle (6) is a sufficient and necessary condition of $BC = XY$.

If $\beta = \gamma$, $BC = XY$ holds only for an equilateral triangle. The Euler line is not defined, but $\tan \beta = \tan \gamma = \tan 60° = \sqrt{3}$.

In our previous solution we pointed out that for right triangles the problem is not defined. Assume that ABC is obtuse. The obtuse angle cannot be at A as in this case X and Y are inner points of BC and $XY < BC$. Let β be obtuse. (See *Figure 1980/1.2*); O, the point of intersection of the perpendicular

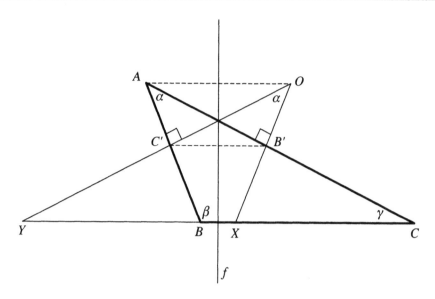

Figure 80/1.2

bisectors is separated from B by AC. and so $\angle BAC = \angle YOX = \alpha$, because the corresponding lines are parallel. If, again, we assume that $XY = BC$, then the altitudes in ABC and OXY at BC and XY have the same length. Indeed, B' and C' are midpoints of the sides of both triangles, BC and XY are on the same line, hence the altitudes of $AB'C'$ and $OC'B'$ at A and O have the same length. As two triangles are congruent if they have a side in common, the opposite angle is the same and the appropriate two altitudes have the same length, moreover they are the reflections of each other to the common perpendicular bisector, f, the triangles $AB'C'$ and $OC'B'$ are the reflections of one another and the same holds for ABC and OXY.

Since $\angle OXY = \beta$, in the triangle $XB'C$ we have $\beta = 90° + \gamma$ so

$$\gamma = -(90° - \beta),$$

and $\tan \gamma = - \tan(90° - \beta) = - \cot \beta$, thus

(7) $\tan \beta \tan \gamma = -1.$

Our arguments can be reversed, thus for an obtuse triangle $BC = XY$ if and only if (7) holds.

Remark. In the second solution we used the following theorem: The Euler line of the triangle ABC is parallel to BC if and only if

(8) $\tan \beta \tan \gamma = 3 \qquad (\beta \neq \gamma).$

Indeed, let the radius of the circumcircle $R = 1$. If the Euler line is parallel to BC, then the centre of gravity, the centre of the circumcircle and the orthocentre lie on the same side of BC. Therefore β and γ must be acute angles and (8) can only be satisfied by acute angles. So we may assume that β and γ are acute. The distance of the centre of the circumcircle and BC is $R\cos\alpha = \cos\alpha$, the distance of the centre of gravity from BC is equal to the third of the altitude h_a, hence the Euler line is parallel to BC if and only if:

(9) $$3\cos\alpha = h_a.$$

Since $2t = ah_a = bc\sin\gamma$ and the sides $b = 2\sin\beta$, $c = 2\sin\gamma$ and $a = 2\sin\alpha$, equation (9) implies that $3a\cos\alpha = 2t = bc\sin\alpha$, i.e.

$$6\sin\alpha\cos\alpha = 4\sin\alpha\sin\beta\sin\gamma,$$
$$3\cos\alpha = -3\cos(\beta+\gamma) = 2\sin\beta\sin\gamma,$$
$$-3\cos\beta\cos\gamma + 3\sin\beta\sin\gamma = 2\sin\beta\sin\gamma,$$
$$\tan\beta\tan\gamma = 3,$$

Thus (8) holds. Now, as all arguments are reversible, (8) implies that the centre of gravity and the centre of the circumcircle are on the same side of BC and they are of the same distance from it, hence the Euler line is parallel to BC.

1980/2. *Define the numbers $a_0, a_1, \ldots, a_n$ in the following way:*

$$a_0 = \frac{1}{2}, \quad a_{k+1} = a_k + \frac{a_k^2}{n} \quad (n > 1, \ k = 0, 1, \ldots, n-1).$$

Prove that

(1) $$1 - \frac{1}{n} < a_n < 1.$$

Solution. The definition of the a_i-s implies that

(2) $$a_0 < a_1 < \ldots < a_n.$$

Examine the difference of the reciprocal of two consecutive terms:

(3) $$\frac{1}{a_k} - \frac{1}{a_{k+1}} = \frac{a_{k+1} - a_k}{a_k a_{k+1}} = \frac{a_k^2}{na_k\left(a_k + \frac{a_k^2}{n}\right)} = \frac{1}{n + a_k}.$$

Since $a_k < a_n$, if $k < n$, we obtain from (2) and (3) that

$$\frac{1}{n+a_n} < \frac{1}{a_k} - \frac{1}{a_{k+1}} < \frac{1}{n}.$$

Adding these inequalities for $k = 0, 1, \ldots, n-1$ we get:

(4) $$\frac{n}{n+a_n} < \frac{1}{a_0} - \frac{1}{a_n} < 1.$$

The right side of our inequality is equivalent to $2 - \frac{1}{a_n} < 1$, i.e. to the inequality $a_n < 1$ from the proof. The left side of (4) implies that:

$$\frac{1}{a_n} < 2 - \frac{n}{n+a_n}.$$

Using $a_n < 1$ this means that

$$\frac{1}{a_n} < 2 - \frac{n}{n+1} = 1 + \frac{1}{n+1} < 1 + \frac{1}{n-1} = \frac{n}{n-1},$$

or

$$a_n > \frac{n-1}{n} = 1 - \frac{1}{n},$$

and this is the left side of the problem.

1980/3. *Prove that the equation*

(1) $$x^n + 1 = y^{n+1},$$

where n is a positive integer not smaller then 2, has no positive integer solution in x and y for which x and $n+1$ are relatively prime.

Solution. Let us assume to the contrary that (1) has a positive integer solution x, y, where x and $n+1$ are relatively prime. Rewrite (1) in the following form:

(2) $$x^n = y^{n+1} - 1 = (y-1)\left(y^n + y^{n-1} + \ldots + y + 1\right).$$

We show that $y-1$ and $y^n + y^{n-1} + \ldots + y + 1$ are coprime; if they are not, then there is a prime p and integers A and B such that

(3) $$y - 1 = Ap, \qquad y = Ap + 1;$$

(4) $$y^n + y^{n-1} + \ldots + y + 1 = Bp$$

holds. Substituting (3) to (4) the terms on the l.h.s give residue 1 mod p, hence there is an integer C such that the l.h.s. of (4) is of the form $Cp + (n+1)$, that is

$$(B - C)p = n + 1.$$

Then p divides $n+1$ and by (2) it divides x, contradicting the assumption that x and $n+1$ are relatively prime.

Thus $y-1$ and $y^n + y^{n-1} + \ldots + y + 1$ are relatively prime. As their product is a full n-th power, both of them are full n-th powers. But, this is impossible as

$$y^n < y^n + y^{n-1} + \ldots + y + 1 < (y+1)^n$$

and there is no full n-th power between the n-th powers of two consecutive integers. This is a contradiction, so we proved our statement.

1980/4. *Determine all positive integers n such that the following statement holds: In the inscribed convex polygon $A_1 A_2 \ldots A_{2n}$ if the pairs of opposite sides*

$$(A_1 A_2, A_{n+1} A_{n+2}), (A_2 A_3, A_{n+2} A_{n+3}), \ldots, (A_{n-1} A_n, A_{2n-1} A_{2n})$$

are parallel, then the sides

$$A_n A_{n+1}, A_{2n} A_1$$

are parallel as well.

First solution. First, we prove that the statement holds for even n-s, but is false in general for odd n-s.

We precede our proof with two remarks. First: If A_1, A_2, A_3 and B_1, B_2, B_3 are consecutive points of a circle such that the two triples have the same orientation, moreover $A_1A_2 \parallel B_1B_2$ and $A_2A_3 \parallel B_2B_3$, then the arc A_1A_3 containing A_2 is equal to the arc B_1B_3 containing B_2 (see *Figure 1980/4.1*).

The angles $\angle A_1A_2A_3 = \angle B_1B_2B_3$, because their sides are parallel and so the angles A_1A_3 and B_1B_3 are measured by the same arc hence they are equal.

Second: if the arcs AB and CD of a circle are of the same length and orientation, then the chords AD and BC are parallel (see *Figure 1980/4.2*). The

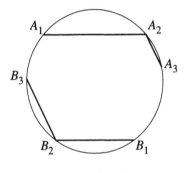

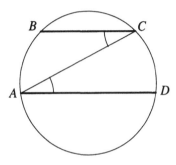

Figure 80/4.1   Figure 80/4.1

condition implies that $\angle ACB = \angle CAD$, because they are measured by he same arc, hence $AD \parallel BC$.

Now, assume that in the convex polygon

$$A_1A_2 \parallel A_{n+1}A_{n+2}, \quad A_2A_3 \parallel A_{n+2}A_{n+3}, \quad \ldots, \quad A_{n-1}A_n \parallel A_{2n-1}A_{2n}.$$

The convexity implies that the triples of points $A_1A_2A_3$ and $A_{n+1}A_{n+2}A_{n+3}$, the triples $A_3A_4A_5$ and $A_{n+3}A_{n+4}A_{n+5}$, $\ldots$, moreover the triples $A_{n-2}A_{n-1}A_n$ $A_{2n-2}A_{2n-1}A_{2n}$ are oriented the same way, hence by our first remark $\overarc{A_1A_3} = \overarc{A_{n+1}A_{n+3}}$, $\overarc{A_3A_5} = \overarc{A_{n+3}A_{n+5}}$, $\ldots$, $\overarc{A_{n-2}A_n} = \overarc{A_{2n-2}A_{2n}}$. Since n is odd, we listed all subarcs of $\overarc{A_1A_n}$ and $\overarc{A_{n+1}A_{2n}}$, hence $\overarc{A_1A_n} = \overarc{A_{n+1}A_{2n}}$. Now, our second remark implies that the sides A_1A_{2n} and A_nA_{n+1} are parallel as well (see *Figure 1980/4.3*).

We show a counterexample for even n-s. We show that the parallelity of $n-1$ pairs of sides does not imply that the remaining two sides are parallel.

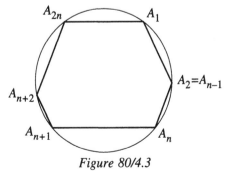

Figure 80/4.3

Choose the vertices of the polygon such a way that for the consecutive pairs of sides the following holds:

$$\overline{A_1A_2} = \overline{A_3A_4} = \ldots = \overline{A_{n-1}A_n} = \frac{3\pi}{2n} = \overline{A_{n+2}A_{n+3}} = \overline{A_{n+4}A_{n+5}} = \ldots = \overline{A_{2n-2}A_{2n-1}};$$

$$\overline{A_2A_3} = \overline{A_4A_5} = \ldots = \overline{A_{n-2}A_{n-1}} = \frac{\pi}{2n} = \overline{A_{n+1}A_{n+2}} = \overline{A_{n+3}A_{n+4}} = \ldots = \overline{A_{2n-1}A_{2n}};$$

$$\overline{A_nA_{n+1}} = \frac{\pi}{n} = \overline{A_{2n}A_1}.$$

Visually, we have arcs of length $\frac{3\pi}{2n}$ (type I.) and $\frac{\pi}{2n}$ (type II.) alternating between the vertices A_1 and A_n and the vertices A_{n+1} and A_{2n} (see *Figure 1980/4.4*, where $n=4$), and so our vertices are among the vertices of a regular $4n$-gon. Now, A_kA_{k+1} and $A_{n+k}A_{n+k+1}$ are parallel for $k \neq n$, as by the arrangement $A_{k+1}A_{n+k} = A_{n+k+1}A_k$ holds. On the other hand the parallelity of A_nA_{n+1} and $A_{2n}A_1$ would imply $A_1A_n = A_{n+1}A_{2n}$. This is false, since A_1A_n consists of $\frac{n}{2}$ edges of type I. and $\frac{n-2}{2}$ edges of type II., hence

$$\overline{A_1A_n} = \frac{n}{2} \cdot \frac{3\pi}{2n} + \frac{n-2}{2} \cdot \frac{\pi}{2n} = \frac{2n-1}{2n}\pi.$$

In $\overline{A_{n+1}A_{2n}}$ we have $\frac{n-2}{2}$ edges of type I. and $\frac{n}{2}$ edges of type II., thus

$$\overline{A_{n+1}A_{2n}} = \frac{n-2}{2} \cdot \frac{3\pi}{2n} + \frac{n}{2} \cdot \frac{\pi}{2n} = \frac{2n-3}{2n}\pi,$$

so their length is different and the two sides are not parallel as we stated.

Second solution. Here we provide a simple proof for the first part using complex numbers. Let a, b, c, d be points on a circle about the origin. It is well known that the lines through a and b, and through c and d are parallel if and only if $ab = cd$.

Consider the polygon $A_1A_2 \ldots A_{2n}$ on the complex plane such that the centre of the circumcircle is the point O. Denoting the numbers of the vertices by the corresponding lower case letters, for odd n the conditions reread as:

$$a_1a_2 = a_{n+1}a_{n+2},$$
$$a_{n+2}a_{n+3} = a_2a_3,$$
$$a_3a_4 = a_{n+3}a_{n+4},$$
$$\vdots$$
$$a_{2n-1}a_{2n} = a_{n-1}a_n.$$

Multiplying the equations and simplifying the products we obtain:

$$a_1a_{2n} = a_na_{n+1},$$

which means that the sides A_1A_{2n} and A_nA_{n+1} are parallel.

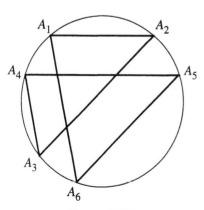

Figure 80/4.5

This proof did not use the convexity of the polygon, so the statement holds for self-intersecting polygons, too. For such an example see *Figure 1980/4.5.*

Remark. The counterexample of the first solution can be formulated with complex numbers as well, but it requires some extra discussion.

1980/5. *In a rectangular coordinate system we call a line parallel to the x axis triangular if it intersects the curve with equation*

$$y = x^4 + px^3 + qx^2 + rx + s$$

in the points A, B, C and D (from left to right) such that the segments AB, AC and AD are the sides of a triangle.

Prove that the lines parallel to the x axis intersecting the curve in four distinct points are all triangular or none of them is triangular.

First solution. Let the line e parallel to x intersect the curve at the points A, B, C, D, moreover let the coordinates of A be $A(t, d)$. Translate A to the origin. The equation of the curve in the new system is given by the substitutions $x \rightarrow x + t$, $y \rightarrow y + d$:

$$y + d = (x + t)^4 + p(x + t)^3 + q(x + t)^2 + r(x + t) + s,$$

hence

$$y = x^4 + (4t + p)x^3 + (6t^2 + 3pt + q)x^2 +$$
$$+ (4t^3 + 3pt^2 + 2tq + r)x + (t^4 + pt^3 + qt^2 + rt + s) - d.$$

Let denote the new coefficients by P, Q, R and W, the equation rereads as:

$$y = x^4 + Px^3 + Qx^2 + Rx + S.$$

As the origin is on the curve, $S = 0$. The x values corresponding to the points A, B, C, D are the roots of the equation

$$x^4 + Px^3 + Qx^2 + Rx = 0$$

and A belongs to $x = 0$. Let denote the x coordinates of B, C, D by t_1, t_2, t_3 ($t_1 < t_2 < t_3$) respectively. These satisfy the equation

$$x^3 + Px^2 + Qx + R = 0.$$

A triangle can be constructed from the segments of length t_1, t_2, t_3 if and only if the inequalities $-t_1 + t_2 + t_3 > 0$, $t_1 - t_2 + t_3 > 0$, $t_1 + t_2 - t_3 >$ hold. The first two are obviously true, hence the condition is equivalent to the inequality:

(8) $\qquad (-t_1 + t_2 + t_3)(t_1 - t_2 + t_3)(t_1 + t_2 - t_3) > 0.$

The connection between the roots and the coefficients of a polynomial say that $t_1 + t_2 + t_3 = -P$, $t_1 t_2 + t_2 t_3 + t_3 t_1 = Q$ and $t_1 t_2 t_3 = -R$. It is easy to verify by calculations that the l.h.s of (8) rereads as $P^3 - 4PQ + 8R$, thus the sufficient and necessary condition for e to be triangular is

$$\delta = P^3 - 4PQ + 8R > 0.$$

Easy calculation shows that

$$\delta = P^3 - 4PQ + 8R = p^3 - 4pq + 8r,$$

hence δ does not depend on the choice of e. Thus if for a line e the number $\delta > 0$, then it holds for every line; and if $\delta \leq 0$, then this holds for all of them. Hence all lines are triangular at the same time or none of them is triangular.

Second solution. A fairly simple solution can be given using visual arguments. Observe on *Figure 1980/5.1* that as $AB < AC < AD$, the condition for the points A, B, C, D being triangular is

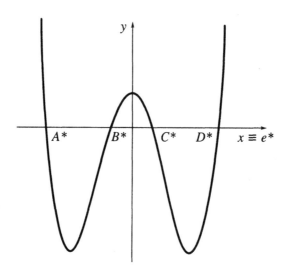

Figure 80/5.1

$$AD < AB + AC, \quad \text{i.e.} \quad AB + BC + CD < AB + AB + BC$$

which implies $CD < AB$. The 4–tuple of points is not triangular if $CD \geq AB$. Now, assume that there is a line e such that $CD < AB$, and another line e' such that for the points of intersection $C'D' \geq A'B'$.

Translating e towards e' there will be a line e^* such that for the points of intersection A^*, B^*, C^*, D^* the equation $C^*D^* = A^*B^*$ holds. Now, choose the coordinate system such that x coincides with e^* and the midpoint of BC is the origin. Denote by $-a$, $-b$, b, a the x coordinates of the points A^*, B^*, C^*, D^* respectively. The polynomial (with leading coefficient 1) is uniquely determined by its roots, the equation rewrites as:

$$y = (x - a)(x + a)(x - b)(x + b) = \left(x^2 - a^2\right)\left(x^2 - b^2\right),$$

hence the curve is symmetric to the y axis.

But then the intersections of a line parallel to x are symmetric, too, hence for every intersecting line e $AB=CD$ contradicting our assumptions.

1980/6. *Find the digits left and right of the decimal point in the decimal form of the number*

$$\left(\sqrt{2}+\sqrt{3}\right)^{1980}.$$

First solution. The examined number changes only a "little bit" if we add its reciprocal, $\left(\sqrt{3}-\sqrt{2}\right)^{1980}$. Set

$$A=\left(\sqrt{3}+\sqrt{2}\right)^{1980}+\left(\sqrt{3}-\sqrt{2}\right)^{1980}.$$

Now,

$$\left(\sqrt{3}+\sqrt{2}\right)^2=5+2\sqrt{6}\quad\text{and}\quad\left(\sqrt{3}-\sqrt{2}\right)^2=5-2\sqrt{6},$$

hence

$$A=\left(5+2\sqrt{6}\right)^{990}+\left(5-2\sqrt{6}\right)^{990}.$$

Using the binomial theorem we obtain:

$$A=5^{990}+\binom{990}{1}5^{989}\cdot2\sqrt{6}+\ldots+\binom{990}{989}5\cdot(2\sqrt{6})^{989}+5^{990}-$$

$$-\binom{990}{1}5^{989}\cdot2\sqrt{6}+\ldots-\binom{990}{989}5\cdot(2\sqrt{6})^{989}+(2\sqrt{6})^{990}=$$

$$=2\cdot5^{990}+\ldots+2\binom{990}{988}5^2\cdot2^{988}\cdot6^{494}+2\cdot2^{990}\cdot6^{495}.$$

Here, every term is an integer, hence A is an integer and all terms except the last one is divisible by 10, hence the last digit of A equals the last digit of the last term. The last term is $2^{991}\cdot6^{495}$. We know that the powers of 6 end by 6 and the powers of 2 end by 2, 4, 8, 6, $\ldots$, periodically. As $991=4\cdot247+3$, the last decimal digit of 2^{991} is 8 and so the last digit of A is the same as the last digit of $8\cdot6$, therefore it equals 8.

On the other hand, $5-2\sqrt{6}<0{,}2$, hence $\left(5-2\sqrt{6}\right)^{990}<(0{,}2)^{990}=$
$=0.4^{495}<0.1^{495}$, hence in the decimal form of $\left(5-2\sqrt{6}\right)^{990}$ there are at least 495 digits that are 0, thus

$$\left(\sqrt{3}+\sqrt{2}\right)^{1980}=\left(5+2\sqrt{6}\right)^{990}=A-\left(\sqrt{3}-\sqrt{2}\right)^{1980}=$$

$$=\overline{XXX\ldots8}-0{,}00\ldots0XX=\overline{XXX\ldots7{,}999\ldots XX\ldots}$$

and the digits before and behind the decimal point are $\ldots7.9\ldots.$.

Second solution. Along the lines of the previous solution let us examine the sequence

$$A_n = \left(5 + 2\sqrt{6}\right)^n + \left(5 - 2\sqrt{6}\right)^n.$$

It is easy to observe that the elements of this series satisfy the equation

$$A_n + A_{n+2} = 10A_{n+1}.$$

Thus the sum of the second consecutive terms of the sequence, is divisible by 10. Since $A_1 = 10$, $A_2 = 98$, the last decimal digits of the terms are

$$0, \ 8, \ 0, \ 2, \ 0, \ 8, \ 0, \ 2, \ \ldots$$

The sequence has period 4. As $990 = 247 \cdot 4 + 2$, the number A_{990} ends by 8. As seen in the first solution, in $\left(5 - 2\sqrt{6}\right)^{990}$ the digit 0 stands after the decimal point. Now

$$\left(\sqrt{2} + \sqrt{3}\right)^{1980} = \left(5 + 2\sqrt{6}\right)^{990} = A_{990} - \left(5 - 2\sqrt{6}\right)^{990}$$

implies that the digits by the decimal point in $\left(\sqrt{2} + \sqrt{3}\right)^{1980}$ are: $\ldots 7, 9 \ldots$.

1981.

1981/1. *Let P be a point inside the triangle ABC and D, E, F are the feet of the perpendiculars from P to the lines BC, CA, AB, respectively. Find all P which minimise:*

(1)
$$\frac{BC}{PD} + \frac{CA}{PE} + \frac{AB}{PF}.$$

First solution. Use the notations of *Figure 1981/1.1*, let A denote the area of the triangle and S the sum in (1).

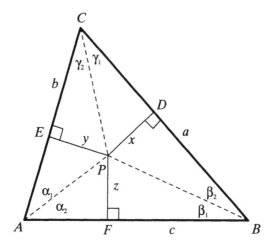

Figure 81/1.1

A is the sum of the areas of the triangles APB, BPC and CPA, hence

$$2T = ax + by + cz,$$

$$S = \frac{a}{x} + \frac{b}{y} + \frac{c}{z}.$$

S is minimal if $2SA$ is minimal, because A does not depend on the choice of P. Calculating $2ST$ and using that the sum of a real number and its reciprocal is at least 2, we get:

$$2ST = \left(\frac{a}{x} + \frac{b}{y} + \frac{c}{z}\right)(ax + by + cz) = a^2 + b^2 + c^2 + ab\left(\frac{x}{y} + \frac{y}{x}\right) +$$

$$+ bc\left(\frac{y}{z} + \frac{z}{y}\right) + ac\left(\frac{x}{z} + \frac{z}{x}\right) \geq a^2 + b^2 + c^2 + 2ab + 2bc + 2ca =$$

$$= (a+b+c)^2,$$

and equation holds if and only if $x = y = z$, thus P is the centre of the incircle. Thus the minimal value of S is:

$$\min S = \frac{(a+b+c)^2}{2T},$$

and S attains it when p is the centre of the incircle.

Second solution. Use the notations of the first solution. The perpendiculars and the lines through P together with the vertices of the triangle determine six right triangles. From the triangles PAF, and PBF we have $AF = z \cot \alpha_2$, and $FB = z \cot \beta_1$, hence as $AF + FB = c$,

$$\frac{c}{z} = \cot \alpha_2 + \cot \beta_1$$

follows. Similarly,

$$\frac{a}{x} = \cot \beta_2 + \cot \gamma_1,$$

$$\frac{b}{y} = \cot \gamma_2 + \cot \alpha_1.$$

From the sum of the latter three inequalities we get:

(2) $\qquad S = (\cot \alpha_1 + \cot \alpha_2) + (\cot \beta_1 + \cot \beta_2) + (\cot \gamma_1 + \cot \gamma_2).$

Now, let us examine the value of the expression in the first parenthesis. As $\alpha_2 = \alpha - \alpha_1$

(3) $\qquad \cot \alpha_1 + \cot \alpha_2 = \frac{\cos \alpha_1}{\sin \alpha_1} + \frac{\cos(\alpha - \alpha_1)}{\sin(\alpha - \alpha_1)} = \frac{\sin \alpha}{\sin \alpha_1 \sin(\alpha - \alpha_1)}.$

Since $\sin \alpha_1$ and $\sin(\alpha - \alpha_1)$ are positive, $4xy \leq (x+y)^2$ implies that

$$\sin \alpha_1 \sin(\alpha - \alpha_1) \leq \left(\frac{\sin \alpha_1 + \sin(\alpha - \alpha_1)}{2}\right)^2 = \sin^2 \frac{\alpha}{2} \cos^2 \frac{\alpha - 2\alpha_1}{2} \leq \sin^2 \frac{\alpha}{2},$$

and equality holds if and only if $\alpha = 2\alpha_1$, that is, when PA is the bisector. Applying it to (3) we get

$$\cot \alpha_1 + \cot \alpha_2 = \frac{\sin \alpha}{\sin^2 \frac{\alpha}{2}} = \frac{2 \sin \frac{\alpha}{2} \cos \frac{\alpha}{2}}{\sin^2 \frac{\alpha}{2}} = 2 \cot \frac{\alpha}{2}.$$

We get similar results for the other two parentheses in (2), thus

$$S \geq 2 \left(\cot \frac{\alpha}{2} + \cot \frac{\beta}{2} + \cot \frac{\gamma}{2} \right),$$

and equality holds if and only if PA, PB, PC are bisectors, hence P is the centre of the incircle.

Remark. In our second solution the equation $c = z \cot \alpha_2 + z \cot \beta_1$ holds even if the angle β_1 is obtuse, and the length of c is the difference of the length of two sides, as $\cot \beta_1 < 0$.

1981/2. *Consider all subsets of size r of the set $H_n = \{1, 2, \ldots, n\}$, where $1 \leq r \leq n$. Each subset has a minimal element, let $F(n,r)$ denote the arithmetic mean of these elements. Prove that*

$$F(n,r) = \frac{n+1}{r+1}.$$

First solution. Let $S(n,r)$ denote the sum of the smallest elements from the r element subsets of H_n. We prove that for $n > 1$ and $r > 1$

(1) $S(n+1,r) = S(n, r-1) + S(n,r).$

Let us order the elements of every subset in increasing order. We can split the r element subsets of H_{n+1} into two disjoint groups as it contains the number $n+1$ or it does not. If we omit $n+1$ from the subsets of the first group, we get the set of $r-1$ element subsets of H_n, and the second group is the set of r element subsets of H_n. Thus (1) holds.

It is easy to observe that this formula is the same that holds for the binomial coefficients. We prove by induction that

(2) $S(n,r) = \binom{n+1}{r+1}.$

If $r = 1$, the subsets of H_n are the sets $\{1\}$, $\{2\}$, $\ldots$, $\{n\}$ hence

$$S(n, 1) = \frac{n(n+1)}{2} = \binom{n+1}{2} = \binom{n+1}{r+1},$$

and so (2) holds in case of $r = 1$ for every n and so, in case $n = 1$, too. It holds in case of $n = 2$, $r = 2$, as $H_2 = \{1, 2\}$ is the unique two element subset of itself, thus $S(2,2)=1$, as in (2). Now, assume the (2) holds for until some n in case $1<r\le n$. Combining (1) with the formula for the binomial coefficients gives that

$$(3) \qquad S(n+1,r)=\binom{n+1}{r}+\binom{n+1}{r+1}=\binom{n+2}{r+1},$$

thus (2) holds under our assumptions.

As there are $\binom{n}{r}$ r-element subsets of H_n,

$$(4) \qquad F(n,r)=\frac{\binom{n+1}{r+1}}{\binom{n}{r}}=\frac{(n+1)!}{(r+1)!(n-r)!}:\frac{n!}{r!(n-r)!}=\frac{n+1}{r+1},$$

so we proved the statement.

Second solution. Let $S(n,r)$ denote the sum of the smallest elements from the r element subsets of H_n. Assume that the elements of the subsets are in increasing order. The possible values of the first elements are: $1, 2, \ldots, n-r+$ $+1$.

The positive integer k is the smallest element of a subset, if the elements were chosen from the elements $k, k+1, \ldots, n$ but we chose k along with $r-1$ other elements from the $n-k$ elements left. The sum of the first elements for these subsets is $k\binom{n-k}{r-1}$, and so

$$S(n,r)=\sum_{k=1}^{n-r+1}k\binom{n-k}{r-1}.$$

This sum is
$$S(n,r)=\binom{n-1}{r-1}+\binom{n-2}{r-1}+\binom{n-3}{r-1}+\ldots+\binom{r}{r-1}+\binom{r-1}{r-1}+$$
$$+\binom{n-2}{r-1}+\binom{n-3}{r-1}+\ldots+\binom{r}{r-1}+\binom{r-1}{r-1}+$$
$$+\binom{n-3}{r-1}+\ldots+\binom{r}{r-1}+\binom{r-1}{r-1}+$$
$$+\ldots\ldots\ldots\ldots\ldots\ldots\ldots\ldots\ldots+$$
$$+\binom{r-1}{r-1}.$$

Now, adding up the sum by the rows, using

(5)
$$\sum_{i=1}^{n-k} \binom{n-i}{k} = \binom{n}{k+1}$$

gives

$$S(n,r) = \binom{n}{r} + \binom{n-1}{r} + \binom{n-2}{r} + \ldots + \binom{r+1}{r} + \binom{r}{r} = \binom{n+1}{r+1}.$$

Since the number of the subsets is $\binom{n}{r}$, the arithmetic mean, (as in the first solution) is:

$$F(n,r) = \frac{\binom{n+1}{r+1}}{\binom{n}{r}} = \frac{n+1}{r+1},$$

that we wanted to prove.

Third solution. Let R_r denote the r element subsets of H_n and R'_{r+1} the $r+1$ element subsets of $H'_n = \{0, 1, 2, \ldots, n\}$.

Let us assume that k is the smallest number in an element of R_r ($k=1$, 2, ...). If we add one of the elements $0, 1, 2, \ldots, k-1$ to the set, then we get an element of R'_{r+1}, and we can get exactly k elements of R'_{r+1} doing this way. Obviously, every element of R'_{r+1} can be obtained in a unique way from an element of R_r (by omitting the smallest element of it). This means that the number of the elements R'_{r+1} equals the sum of the smallest elements of the elements of R_r.

So we determine the required arithmetic value on the following way:

$$F(n,r) = \frac{\text{the number of elements of } R'_{r+1}}{\text{the number of elements of } R_r} = \binom{n+1}{r+1} : \binom{n}{r}$$

Hence by (1)

$$F(n,r) = \frac{n+1}{r+1}.$$

Remark. In our first solution we used the formulas (3) and (5) of the binomial coefficients (see their proofs in [28]).

2. With a slight modification of our arguments we can prove that the arithmetic mean of the largest elements is

$$r \frac{n+1}{r+1}.$$

1981/3.   *Determine the maximum value of $m^2 + n^2$ where m and n are integers satisfying $m, n \in \{1, 2, \ldots, 1981\}$ and*

$$(1) \qquad\qquad (n^2 - nm - m^2)^2 = 1.$$

Solution. First, $m > n$ cannot be the case as it implies $m^2 > n^2$, $n^2 - m^2 < -1$, $-nm < -1$, $n^2 - nm - m^2 < -2$, contradicting (1).

If $n = m$, the only solution of (1) is $m = n = 1$.

In case $m < n$, $n - m$ is a positive integer. We show that if (n, m) is a solution of (1), then $(m, n - m)$ is a solution of (1), too:

$$\left(m^2 - m(n - m) - (n - m)^2\right)^2 = \left(-(n^2 - nm - m^2)\right)^2 = 1.$$

This means that if (n, m) is a solution, then $m < n$ and the pair (n', m') is a solution, too, where

$$(2) \qquad\qquad \begin{aligned} n' &= m \\ m' &= n - m. \end{aligned}$$

Reversing (2) we get that if (n', m') is a solution, where $m' < n'$, then (n, m) is a solution, too, where

$$(3) \qquad\qquad \begin{aligned} n &= m' + n' \\ m &= n'. \end{aligned}$$

(3) implies that starting with the pair $(1, 1)$ we get infinitely many solutions:

$$(4) \qquad (1, 1),\ (2, 1),\ (3, 2),\ (5, 3),\ \ldots,\ (1597, 987),\ (2584, 1597),\ \ldots$$

We show that in (4) we listed all solutions of (1). Indeed, let (n, m) be a solution; if $m < n$, then by (2) we can find a "smaller" solution (n', m'). The procedure terminates at a pair, where $m' = n'$; but this has to be the pair $n' = m' = 1$. As a pair determines all the solutions given by (2) and (3), (4) provides all the solutions. As in (4) the pair $(1597, 987)$ is the largest not exceeding 1981, the maximum of $n^2 + m^2$ is:

$$1597^2 + 987^2 = 3\,524\,578.$$

Remark. (3) implies that the numbers occurring in the solutions are the Fibonacci numbers defined by $a_0 = 1$, $a_1 = 1$, $a_n = a_{n-1} + a_{n-2}$. Thus the pairs are

$$(a_1, a_0),\ (a_2, a_1),\ (a_3, a_2),\ \ldots,\ (a_n, a_{n-1}),\ \ldots$$

1981/4. a) *For which $n > 2$ is there a set of n consecutive positive integers such that the largest number in the set is a divisor of the least common multiple of the remaining $n - 1$ numbers?*

b) *For which $n > 2$ is there exactly one set having this property?*

First solution. Let first $n = 3$ and the first three consecutive integers be: $s - 2$, $s - 1$, s. Since s and $s - 1$ are coprime, s can divide $(s - 2)(s - 1)$ only if it divides $s - 2$, hence $s = 1$ or 2, but then the 3 numbers are not all positive, thus for $n = 3$ the requested numbers do not exist.

Now, let $n = 4$, and the four consecutive numbers: $s - 3$, $s - 2$, $s - 1$, s. If s divides the least common multiple of the other three numbers, then it divides their product. It is coprime to $s - 1$ so it divides $(s - 3)(s - 2)$. As

$$(s - 3)(s - 2) = s(s - 5) + 6,$$

s divides 6. $s \neq 1$, 2, 3 because then 0 were among the 4 numbers. Hence $s = 6$. Then the 4-tuple is: 3, 4, 5, 6, and 6 divides 60, the least common multiple of the other three numbers.

Finally, let $n \geq 5$. Choose s to be $s = (n - 1)(n - 2)$ and let $k = \dfrac{s}{2}$. We shall provide two n-tuples of consecutive integers:

α) $s - n + 1$, $s - n + 2$, $\ldots$, $s - 1$, s;

β) $k - n + 1$, $k - n + 2$, $\ldots$, $k - 1$, k.

These numbers are positive: In case α) $s - n + 1 = (n - 1)(n - 2) - n + 1 = (n - 3)(n - 1) \geq 8$ since $n \geq 5$.

In case β)

$$k - n + 1 = \frac{(n - 1)(n - 2)}{2} - n + 1 = \frac{n(n - 5)}{2} + 2 \geq 2.$$

As α) and β) are n consecutive numbers, they contain numbers divisible by $n - 1$ and $n - 2$. As $n - 1$ and $n - 2$ are coprime, their least common multiple is divisible by $(n - 1)(n - 2)$, hence by s and k, as well.

In conclusion there is no solution for $n = 3$, there is a unique solution for $n = 4$, and for $n \geq 5$ there are at least two solutions.

Second solution. Let s denote the largest of the n consecutive numbers.

Let $n = 3$; Now, our numbers are: $s - 2$, $s - 1$, s. If p is a prime divisor of s, then (since s and $s - 1$ are coprime), p divides $s - 2$, and hence p divides 2, too. Thus $p = 2$. The power of p in s is at most 1, because in other case 4 does not divide $s - 2$, thus the only possibility is $s = 2$, but then, $s - 2 = 0$ would hold and so there is no solution for $n = 3$.

Let $n = 4$; then $s > 3$, and our numbers are: $s - 3$, $s - 2$, $s - 1$, s. A prime divisor p of s can only divide $s - 2$ or $s - 3$ hence p is 2 or 3 and they can only occur on at most the first power in s. The only possible choice for s is $2 \cdot 3 = 6$. And, indeed, in the 4-tuple 3, 4, 5, 6 we have that 6 divides $3 \cdot 4 \cdot 5$.

Now, let $n \geq 5$. Let us put n between two powers of 2, thus let r be a positive integer such that

$$2^r < n \leq 2^{r+1}.$$

Choose $s = 3 \cdot 2^r$ or $s = 5 \cdot 2^r$, the n consecutive integers are:

(1) $s + 1 - n, \quad s + 2 - n, \quad \ldots, \quad s - 1, \quad s.$

The lists in (1) consist of positive integers as, e.g., in the first case $n \leq 2 \cdot 2^r <$ $< 3 \cdot 2^r = s$, $s - n > 0$, and hence $s + 1 - n$ is a positive integer.

As $n > 2^r$ and $n > 3$, the integers $s - 2^r$ and $s - 3$ are in (1) if $s = 3 \cdot 2^r$. As 2^r and 3 are coprime, the least common multiple of the first $n - 1$ numbers in the first n-tuple is divisible by $3 \cdot 2^r = s$. Similarly, if $s = 5 \cdot 2^r$, then for $n \geq 5$ the least common multiple is divisible by $5 \cdot 2^r = s$.

Summarizing: if $n = 3$ then there is no solution, in case $n = 4$ there is a unique solution, and for $n \geq 5$ there are at least two solutions.

1981/5. *Three circles of equal radii have a common point O and lie inside a given triangle. Each circle touches a pair of sides of the triangle. Prove that the incentre and the circumcentre of the triangle are collinear with the point O.*

Solution. Let A, B, C denote the vertices of the triangle and A', B', C' the centre of the circles touching the pair of sides originating from the corresponding vertex. As these centres are of the same distance from 2-2 sides, they lie on the appropriate bisectors and the sides of the triangles ABC and $A'B'C'$ are parallel. (See *Figure 1981/5.1*). Consequently, the triangles ABC and $A'B'C'$

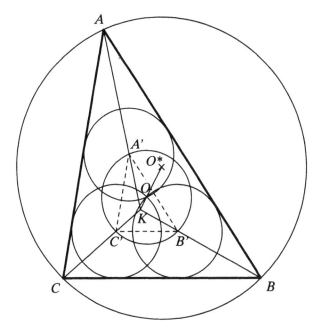

Figure 81/5.1

are homothetic and the centre of the similitude is K, the incentre. O is of the same distance from the vertices A', B', C' hence it is the circumcentre of the triangle $A'B'C'$.

The enlargement with centre K mapping the triangle $A'B'C'$ to the triangle ABC maps O to the circumcentre O^* of the triangle, ABC hence K, O, and O^* are collinear.

1981/6. *The function $f(x,y)$ satisfies*

(1) $$f(0, y) = y + 1,$$

(2) $$f(x+1, 0) = f(x, 1),$$

(3) $$f(x+1, y+1) = f(x, f(x+1, y))$$

for every integer x and y. Find $f(4, 1981)$.

Solution. Let y be an arbitrary positive integer. We want to find the values $f(1, y)$, $f(2, y)$, $f(3, y)$; for simplicity we write over the equation sign the label of the identity we use.

(4) $$f(1, 0) \overset{(2)}{=} f(0, 1) \overset{(1)}{=} 2.$$

$$f(1, y+1) \overset{(3)}{=} f(0, f(1, y)) \overset{(1)}{=} f(1, y) + 1 \overset{(3)}{=}$$
$$\overset{(3)}{=} f(0, f(1, y-1)) + 1 \overset{(1)}{=} f(1, y-1) + 2,$$

thus

$$f(1, y) = f(1, y-1) + 1.$$

Repeating our arguments we obtain:

(5) $$f(1, y) = f(1, 0) + y \overset{(4)}{=} f(0, 1) + y \overset{(4)}{=} y + 2.$$

Now, we determine the value of $f(2, y)$:

$$f(2, y) \overset{(3)}{=} f(1, f(2, y-1)) \overset{(5)}{=} f(2, y-1) + 2.$$

Iterating the procedure we get:

(6) $$f(2, y) = f(2, 0) + 2y \overset{(2)}{=} f(1, 1) + 2y \overset{(5)}{=} 2y + 3.$$

We continue with $f(3, y)$:

$$f(3, y) \overset{(3)}{=} f(2, f(3, y-1)) \overset{(6)}{=} 2f(3, y-1) + 3.$$

Repeating the arguments gives

$$f(3, y) = 2(2f(3, y-2) + 3) + 3 = 2^2 f(3, y-2) + 3 + 2 \cdot 3 = \ldots =$$
$$= 3 + 2 \cdot 3 + 2^2 \cdot 3 + \ldots + 2^{y-1} \cdot 3 + 2^y f(3, 0) \overset{(2)}{=}$$
$$\overset{(2)}{=} 3(2^y - 1) + 2^y f(2, 1) \overset{(6)}{=} 3(2^y - 1) + 2^y \cdot 5,$$

(7) $$f(3,y) = 2^{y+3} - 3.$$

Finally, we determine the value of $f(4,y)$:

$$f(4,y) \overset{(3)}{=} f(3, f(4, y-1)) \overset{(7)}{=} 2^{f(4,y-1)+3} - 3 \overset{(7)}{=}$$

$$\overset{(7)}{=} 2^{\left(2^{f(4,y-2)+3} - 3\right)+3} - 3 = 2^{2^{f(4,y-2)+3}} - 3.$$

Repeating the arguments we get

$$f(4,y) = 2^{2^{2^{\cdot^{\cdot^{\cdot^{2^{f(4,0)+3}}}}}}} - 3,$$

and there are y many 2-s on the r.h.s. As

$$f(4,0) \overset{(2)}{=} f(3,1) \overset{(7)}{=} 2^4 - 3 = 2^{2^2} - 3,$$

$$f(4,y) = 2^{2^{2^{\cdot^{\cdot^{\cdot^{2}}}}}} - 3 \quad \text{(with } y+3 = 1984 \text{ 2-s).}$$

Remark. The function $f(x,y)$ in the problem is a so called doubly recurrence function. The recurrence functions are functions defined on the nonnegative integers attaining nonnegative integer values that are built up from some basic functions with finitely many substitutions.

Our $f(x,y)$ is a well known function introduced by *W. Ackermann* in 1928. This form of the function is due to *R. Péter* (1934).

1982.

1982/1. *The function $f(n)$ is defined on the positive integers and takes on non-negative integer values. For all n, m*

(1) $$f(m+n) - f(m) - f(n) = 0 \quad or \quad 1,$$

(2) $$f(2) = 0, \qquad f(3) > 0,$$

(3) $$f(9999) = 3333.$$

Determine $f(1982)$.

First solution. Let $m = n = 1$, then from (1) it follows that

$$f(2) - 2f(1) = 0 \quad or \quad 1,$$

(2) implies that $2f(1) = 0$ or -1, hence $f(1) = 0$.

Similarly, if $m=2$, $n=1$
$$f(3)-f(2)-f(1)=f(3)=0 \quad \text{or} \quad 1,$$
and by (2)we have $f(3)=1$. Using induction we show that for every positive integer k
$$(4) \qquad f(3k) \geq k.$$
Assume that (4) holds until some k. Let $m=3k$, $n=3$, then (1) implies
$$f(3(k+1))=f(3k+3) \geq f(3k)+f(3) \geq k+1,$$
hence (4) holds for every k. If in the last step $f(3k)>k$ then $f(3(k+1))>k+1$ holds, that is, if for some k_0 we have $f(3k_0)>k_0$ in (4), then for every $k>k_0$ $f(3k)>k$ holds, as well. This observation implies that in case $k<3333$, $f(3k)>$ $>k$ cannot hold as then $f(3\cdot3333)>3333$ follows, contradicting (3); therefore $f(3\cdot1982)=1982$. Now, (1) implies
$$1982=f(3\cdot1982)=f(2\cdot1982+1982) \geq f(2\cdot1982)+f(1982),$$
and
$$f(1982+1982) \geq 2f(1982).$$
Combining the last two results we get:
$$(5) \qquad 1982 \geq 3f(1982), \qquad f(1982) \leq \frac{1982}{3} < 661.$$
Using (1) again
$$f(1982)=f(1980+2) \geq f(1980)+f(2)=f(3\cdot660)+0=660,$$
thus from (5)
$$660 \leq f(1982) < 661,$$
and so
$$f(1982)=660.$$

Second solution. We show that the value of $f(1982)$ can be determined without condition (2). (1) can be rewritten in the following form:
$$(6) \qquad f(m)+f(n) \leq f(m+n) \leq f(m)+f(n)+1.$$
We prove by induction that
$$(7) \qquad xf(y) \leq f(xy) \leq xf(y)+x-1,$$
where x and y are positive integers.

For $x=1$ and y arbitrary the statement is obvious. Assume that (7) holds for $x-1$, that is
$$(8) \qquad (x-1)f(y) \leq f((x-1)y) \leq (x-1)f(y)+x-2.$$
From (6) with substitutions $m=(x-1)y$ and $n=y$
$$(9) \qquad f((x-1)y)+f(y) \leq f(xy) \leq f((x-1)y)+f(y)+1.$$
The left inequality of (8) implies:
$$f((x-1)y)+f(y) \geq (x-1)f(y)+f(y)=xf(y).$$
The right inequality of (8) implies:
$$f((x-1)y)+f(y)+1 \leq (x-1)f(y)+f(y)+1+x-2=xf(y)+x-1.$$
The two latter results combined with (9) gives the required (7).

Now, apply (7) for $x = 1982$, $y = 9999$ and let P denote $f(xy)$:

(10) $$1982 \cdot 3333 \leq P \leq 1982 \cdot 3333 + 1982$$

Now, we substitute $x = 9999$ and $y = 1982$ into (7):

(11) $$9999 \cdot f(1982) \leq P \leq 9999 \cdot f(1982) + 9998$$

Comparing the l.h.s of (10) with the r.h.s of (11) and the r.h.s of (10) with the l.h.s of (11) we get that

$$f(1982) \geq \frac{1982 \cdot 3333 - 9998}{9999} = 659{,}66\ldots ,$$

$$f(1982) \leq \frac{1982 \cdot 3333 + 1982}{9999} = 660{,}86\ldots$$

and since $f(1982)$ is an integer, we obtain

$$f(1982) = 660.$$

1982/2. *A non-isosceles triangle $A_1 A_2 A_3$ has sides a_i opposite to A_i. M_i is the midpoint of side a_i and T_i is the point where the incircle touches side a_i.*

Denote by S_i the reflection of T_i in the interior bisector of angle A_i. Prove that the lines $M_1 S_1$, $M_2 S_2$ and $M_3 S_3$ are concurrent.

Solution. Under the assumptions of the problem, all upcoming points are distinct; Indeed, if e.g. S_1 coincides with S_2, then the central angle belonging to the arc $T_1 T_2$ containing S_1 is twice the angle of the bisectors of the angles A_1 and A_2, $90° + \dfrac{\gamma}{2}$, what is impossible (see *Figure 1982/2.1*).

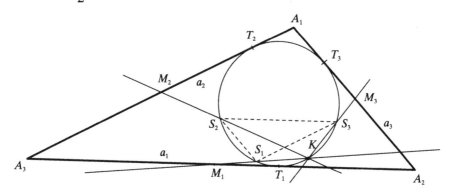

Figure 82/2.1

Since S_1 and T_1, moreover T_3 and T_2 are symmetric to the bisector of A_1, we have $T_1 T_2 = S_1 T_3$; Similarly, $T_1 T_2 = S_2 T_3$; hence $S_1 T_3 = S_2 T_3$.

The tangent of the circumcircle of the isosceles triangle $S_1 S_2 T_3$ (that is the incircle of the triangle $A_1 A_2 A_3$) at the point T_3 and the side $S_1 S_2$ are parallel, so $A_1 A_2$ is parallel to $S_1 S_2$. But the median $M_1 M_2$ and $A_1 A_2$ are parallel too, hence $S_1 S_2$ is parallel to $M_1 M_2$. Similar arguments show that the segments $S_2 S_3$ and $M_2 M_3$, and the segments $S_3 S_1$ and $M_3 M_1$ are parallel. Hence the sides of the triangles $S_1 S_2 S_3$ and $M_1 M_2 M_3$ are pairwise parallel, so the two triangles are similar.

As in this case the circumcircle of $S_1 S_2 S_3$ is smaller than the circle through the midpoints of the sides, the triangle $S_1 S_2 S_3$ is smaller than the triangle $M_1 M_2 M_3$, hence the lines $M_1 S_1$, $M_2 S_2$, $M_3 S_3$ concur in K, the centre of the enlargement, and this is what we wanted to prove.

Remark. The enlargement with centre K maps the two circumcircles, that is, the incircle and the Feuerbach circle to one another. By the theorem of Feuerbach the incircle and the Feuerbach circle are touching, the touching point is at the same time the centre of the enlargement, so K lies on both circles.

This observation simplifies the construction of the touching point of the two circles. Note that the observations can be completed to a proof of Feuerbach's theorem.

1982/3. *Consider the infinite non increasing sequence $\{x_i\}$ of positive reals such that $x_0 = 1$.*

a) *Prove that for every such sequence there is an $n \geq 1$, such that*

(1)
$$S_n = \frac{x_0^2}{x_1} + \frac{x_1^2}{x_2} + \ldots + \frac{x_{n-1}^2}{x_n} \geq 3{,}999.$$

b) *Find such a sequence for which*

$$S_n = \frac{x_0^2}{x_1} + \frac{x_1^2}{x_2} \ldots + \frac{x_{n-1}^2}{x_n} < 4.$$

for all n.

First solution. As $(x_i - 2x_{i+1})^2 \geq 0$, $x_i^2 \geq 4x_i x_{i+1} - 4x_{i+1}^2$ follows, and so

$$\frac{x_i^2}{x_{i+1}} \geq 4(x_i - x_{i+1}).$$

Applying this to the sum in (1):

$$S_n = \frac{x_0^2}{x_1} + \frac{x_1^2}{x_2} + \ldots + \frac{x_{n-1}^2}{x_n} \geq 4(x_0 - x_1 + x_1 - x_2 + \ldots + x_{n-1} - x_n) = 4(1 - x_n).$$

If the limit of the sequence x_n is 0, then there exists an n, such that $x_n \leq \dfrac{1}{4000}$, and for this n

$$S_n \geq 4\left(1 - \frac{1}{4000}\right) = 3{,}999.$$

Now, let the limit c be different from 0, $c > b > 0$, so every element of the sequence is greater than b. Let $n \geq \dfrac{4}{b}$. Then

$$S_n = \sum_{i=0}^{n} \frac{x_i^2}{x_{i+1}} > \sum_{i=0}^{n} \frac{x_i x_{i+1}}{x_{i+1}} = \sum_{i=0}^{n} x_i > nb \geq 4,$$

and the statement is true.

For b) let $x_i = 2^{-i}$ $(i = 0, 1, \ldots)$. Now

$$\frac{x_i^2}{x_{i+1}} = \frac{2^{-2i}}{2^{-(i+1)}} = 2^{1-i}$$

and so

$$S_n = 2 + 1 + \frac{1}{2} + \ldots + \frac{1}{2^{n-1}}.$$

Extending it to an infinite geometric sequence the sum is 4, hence the finite sum S never reaches 4.

Second solution. Introduce the notation

$$a_k = \frac{x_{k-1}}{x_k} \qquad (k = 1, 2, \ldots)$$

here $a_k \geq 1$ and

$$S_n = a_1 + \frac{a_2}{a_1} + \frac{a_3}{a_1 a_2} + \ldots + \frac{a_n}{a_1 a_2 \ldots a_{n-1}} =$$

$$= a_1 + \frac{1}{a_1}\left(a_2 + \frac{1}{a_2}\left(a_3 + \ldots + \frac{1}{a_{n-2}}\left(a_{n-1} + \frac{a_n}{a_{n-1}}\right)\ldots\right)\right).$$

Apply now the inequality $a_k + \dfrac{B}{a_k} \geq 2\sqrt{B}$ that is a consequence of the A.M.–G.M. inequality if $B \geq 0$:

$$S_n \geq 2\sqrt{a_2 + \frac{1}{a_2}\left(a_3 + \ldots + \frac{1}{a_{n-2}}\left(a_{n-1} + \frac{a_n}{a_{n-1}}\right)\ldots\right)} \geq \ldots \geq$$

$$\geq 2\sqrt{2\sqrt{\ldots 2\sqrt{a_{n-1} + \frac{a_n}{a_{n-1}}}}} \geq$$

$$\geq 2\sqrt{2\sqrt{\ldots 2\sqrt{a_n}}} \geq 2\sqrt{2\sqrt{\ldots \sqrt{2}}} = 2^{1 + \frac{1}{2} + \frac{1}{4} + \ldots + \frac{1}{2^{n-1}}}.$$

The limit of the sum in the exponent is 2, so S_n can get arbitrarily close to 4, hence we proved a). (Note that in case $n = 14$ $S_{14} > 3{,}999$ is satisfied.)

See the first solution for the answer for b).

1982/4. *Prove that if n is a positive integer such that the equation*
$$x^3 - 3xy^2 + y^3 = n$$
has a solution in integers x, y, then it has at least three such solutions. Show that the equation has no solutions in integers for n = 2891.

Solution. As
$$x^3 - 3xy^2 + y^3 = (y-x)^3 - 3(y-x)x^2 - x^3,$$
if (x,y) is a solution of our equation, then (x_1, y_1) is a solution, too, where

(1)
$$x_1 = -x + y,$$
$$y_1 = -x.$$

Similarly, with the pair (x_1, y_1), the pair (x_2, y_2) is a solution, as well, where, by (1)
$$x_2 = -x_1 + y_1 = -y,$$
$$y_2 = -x_1 = x - y.$$
The equality of (x,y) and (x_1, y_1) means that $y = 2x$ and $y = -x$, implying $x = y = 0$; hence the transformation in (1) gives a new pair, different from (x,y). (1) gives (x_2, y_2) from (x_1, y_1) and (x,y) from (x_2, y_2), thus the three pairs of solutions are distinct, hence we have at least three solutions.

If $n = 2891$, the equation has no solution. We shall prove it by examining the divisibility of x and y by 3; 2891 gives remainder 2 mod 3 and mod 9. We show that this cannot be the case on the l.h.s.

a) If x and y are of the form $3k$, then the l.h.s. is divisible by 3.

b) If $x = 3k$, $y = 3k \pm 1$, the remainder of the l.h.s mod 3 is ± 1, the same holds if $x = 3k \pm 1$ and $y = 3k$.

c) If $x = 3k + 1$, $y = 3k - 1$ (or in the reverse order), 3 divides $x + y$ and the l.h.s too, as it can be rewritten as $(x+y)(x^2 - xy + y^2) - 3xy^2$.

d) If $x = 3k + 1$, $y = 3k + 1$, the remainder of the l.h.s is 8 mod 9.

e) If $x = 3k - 1$, $y = 3k - 1$, the remainder of the l.h.s is 1 mod 9.

Thus we proved our statement.

1982/5. *The diagonals AC and CE of the regular hexagon ABCDEF are divided by inner points M and N respectively, so that:*

(1)
$$\frac{AM}{AC} = \frac{CN}{CE} = r.$$

Determine r, if B, M and N are collinear.

First solution. Choose the length of the diagonals $AC = CE$ as unit. In this case (1) means that
$$AM = CN = r.$$
With this choice the length of the side of the hexagon and the radius of the circumcircle: $\varrho = \dfrac{1}{\sqrt{3}}$ (see *Figure 1982/5.1*).

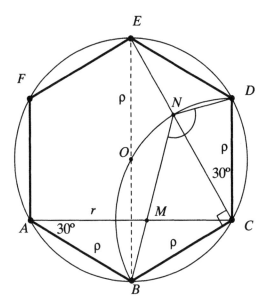

Figure 82/5.1

The rotation around the centre of the hexagon by $120°$ maps the triangle BAC to DCE, the segment BM to DN, so $\angle BND = 120°$. Thus BD is subtended by an angle of $120°$ from N and also from the centre O of the circumcircle of the hexagon. As $\angle BCD = 120°$, the points B, N and D lie on the circle with centre C and radius $BC = CD = \varrho$; thus

$$\varrho = CN = r = \frac{1}{\sqrt{3}}.$$

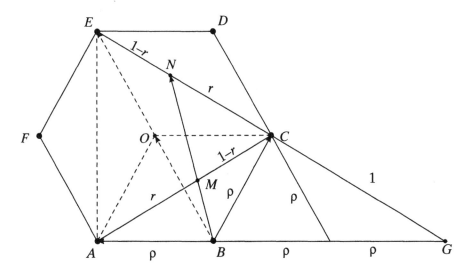

Figure 82/5.2

Second solution. Let the length of the diagonals $AC = CE$ be 1, then the length of the side of the hexagon is: $\varrho = \dfrac{1}{\sqrt{3}}$, and let G denote the intersection of the lines EC and AB. With this choice (1) can be rewritten as (see *Figure 1982/5.2*):

$$AM = CN = r.$$

In the triangle ACG we have $\angle A = 30°$; $\angle C = 120°$ that implies $\angle G = 30°$. Hence it is an isosceles triangle: $AC = CG = 1$, $AG = \sqrt{3} = 3\varrho$.

Apply Menelaus' theorem ([29]) to the triangle ACG and secant BM:

$$1 = \frac{AB}{BG} \cdot \frac{GN}{NC} \cdot \frac{CM}{MA} = \frac{\varrho}{2\varrho} \cdot \frac{1+r}{r} \cdot \frac{1-r}{r},$$

hence

$$3r^2 = 1, \qquad r = \frac{1}{\sqrt{3}} = \varrho.$$

Third solution. Let B be the origin of the coordinate system and denote the vectors by boldface lower case letters corresponding to the points labelled by capital letters (see *Figure 1982/5.2*).

The vector pointing to O is $\mathbf{a} + \mathbf{c}$, hence $\mathbf{e} = 2(\mathbf{a} + \mathbf{c})$. By the conditions:

$$\mathbf{m} = (1-r)\mathbf{a} + r\mathbf{c}, \qquad \mathbf{n} = (1-r)\mathbf{c} + r\mathbf{e} = 2r\mathbf{a} + (1+r)\mathbf{c}.$$

As $\mathbf{m}$ and $\mathbf{n}$ are parallel, there is a real λ such that $\mathbf{n} = \lambda\mathbf{m} = \lambda(1-r)\mathbf{a} + \lambda r\mathbf{c}$. As the decomposition of $\mathbf{n}$ into components parallel to $\mathbf{a}$ and $\mathbf{c}$ is unique, we have

$$\lambda(1-r) = 2r, \qquad \lambda r = 1 + r, \qquad \frac{1-r}{r} = \frac{2r}{1+r}, \qquad 3r^2 = 1, \qquad r = \frac{1}{\sqrt{3}}.$$

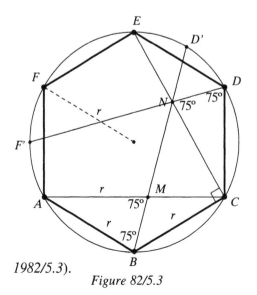

1982/5.3).

Figure 82/5.3

Remark. 1. The third solution shows that we did not use the regularity of the hexagon in full. Indeed, the statement holds for every affine-regular hexagon, the projective image of a regular hexagon.

2. The structure of the problem came from the diagonals of a regular dodecahedron. Its vertices are the midpoints of the arcs over the sides of the hexagon. This shows the background of the problem and suggests other ways to solve it (see *Figure*

1982/6. *Let S be a square with sides length* 100. *Let L be a path within S which does not meet itself and which is composed of line segments A_0A_1, A_1A_2, ..., $A_{n-1}A_n$, where $A_0 \neq A_n$.*

Suppose that for every point P on the boundary of S there is a point of L of distance from P no greater than $1/2$. Prove that there are two points X and Y of L such that the distance between X and Y is not greater than 1 and the length of the part of L which lies between X and Y is not smaller than 198.

Solution. In our solution we shall use that the points of distance more than $\frac{1}{2}$ from the segment RS lie outside of a "stadium-shaped" region (see figure at the solution of problem 1973/4), therefore the points of a line g, or e that are of distance at

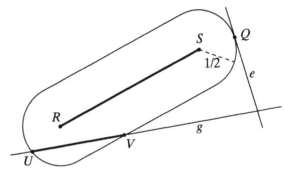

Figure 82/6.1

most $\frac{1}{2}$ from RS are the intersection of the line and the region, that is a close segment or a single point (see *Figure 1982/6.1*).

Assume that starting from A_0 the path L approaches first A from among the vertices of S at a distance not greater than $\frac{1}{2}$. Then, at the point M the path L approaches D at a distance of at least $1/2$, $DM \leq \frac{1}{2}$, before it approaches B. M divides L into two parts, to the polygonal paths A_0M and MA_n (see *Figure 1982/6.2*).

Now, colour red those vertices of the sides of the square that are of distance at most $1/2$ from A_0M. According to our

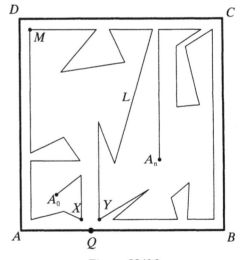

Figure 82/6.2

remark, this is the union of (not necessarily disjoint) closed segments and individual points.

Similarly, colour green the points that are not farther from MA_n than $1/2$. A is red and B is green because by the definition of M, we have that A_0M approaches A but A_0M does not approach B, hence MA_n approaches it.

By our colouring rules and the conditions of the problem, every point of AB is coloured by red or green, and there may be points coloured by both colours. As the monochromatic points consist of closed segments, there is a bicolour point of AB.

Since Q is red, there is a point X on the polygonal path A_0M such that $QX \leq \dfrac{1}{2}$, and as Q is green, too, there is a point Y on the polygonal path MA_n such that $QY \leq \dfrac{1}{2}$. Hence by the triangle inequality:

$$XY \leq QX + QY \leq 1.$$

The distances of the points X, Y from AB, and the distance M from CD is at most $1/2$, hence $XM \geq 99$ and $MY \geq 99$, thus

$$XM \leq XM \text{ polygonal path} ; \qquad MY \leq MY \text{ polygonal path,}$$

and so XMY polygonal path $= XM$ polygonal path $+ MY$ polygonal path $\geq$ $\geq XM + YM \geq 198$, and this is what we wanted to prove.

1983.

1983/1. *Find all functions f defined on the set of positive reals which take positive real values and satisfy:*

(1) $$f(xf(y)) = yf(x)$$

for every positive x and y. Show that

(2) $$f(x) \to 0, \qquad if \quad x \to \infty.$$

Solution. Let the positive number a call a fix point of f if $f(a) = a$. f has a fixpoint: since by (1) $xf(x) = f(xf(x))$, the number $xf(x)$ is a fixpoint. Let b be a fixpoint:

(3) $$f(b) = b.$$

We prove that b^n (n positive integer) is a fixpoint, too. For $n = 1$ this is (3). Now, assume that $f(b^{n-1}) = b^{n-1}$; (1) implies

$$f(b^n) = f(bf(b^{n-1})) = b^{n-1} \cdot f(b) = b^{n-1} \cdot b = b^n;$$

thus b^n is a fixpoint, too. Applying (1), again:

$$b = f(b) = f(1 \cdot b) = f(1 \cdot f(b)) = bf(1),$$

and by $b > 0$

(4) $$f(1) = 1$$

follows, thus 1 is a fixpoint. Hence

$$1 = f(1) = f\left(\frac{1}{b} \cdot b\right) = f\left(\frac{1}{b} \cdot f(b)\right) = bf\left(\frac{1}{b}\right),$$

$$f\left(\frac{1}{b}\right) = \frac{1}{b}.$$

Thus $\frac{1}{b}$ is a fixpoint, and by the preceding arguments $\frac{1}{b^n}$ is a fixpoint, too.

This means that at the points

(5) $b, b^2, b^3, \ldots, b^n, \ldots$

f attains the values

$$b, b^2, b^3, \ldots, b^n, \ldots$$

and at the points

(6) $\frac{1}{b}, \frac{1}{b^2}, \frac{1}{b^3}, \ldots, \frac{1}{b^n}, \ldots$

the function f attains

$$\frac{1}{b}, \frac{1}{b^2}, \frac{1}{b^3}, \ldots, \frac{1}{b^n}, \ldots$$

If $b > 1$, (5) contradicts (2); if $b < 1$, (6) contradicts (2), thus $b = 1$ is the unique fixpoint and so for every positive x

$$x f(x) = 1, \qquad \text{that is} \qquad f(x) = \frac{1}{x}.$$

$f(x) = \frac{1}{x}$ is the only possible solution, and it satisfies the conditions of the problem:

$$f(xf(y)) = f\left(\frac{x}{y}\right) = \frac{y}{x} \qquad \text{and} \qquad yf(x) = y \cdot \frac{1}{x} = \frac{y}{x},$$

and if

$$x \to \infty, \qquad f(x) = \frac{1}{x} \to 0.$$

1983/2. *Let A be one of the two distinct points of intersection of two unequal coplanar circles C_1 and C_2 with centres O_1 and O_2, respectively. One of the common tangents to the circles touches C_1 at P_1, C_2 at P_2 while the other touches C_1 at Q_1 and C_2 at Q_2. Let M_1 be the midpoint of P_1Q_1 and M_2 be the midpoint of P_2Q_2. Prove that $\angle O_1AO_2 = \angle M_1AM_2$.*

First solution. We shall prove that

(1) $\angle O_1AM_1 = \angle O_2AM_2,$

which is equivalent to the statement. Let K denote the point of intersection of the two tangents, this is the centre of similitude of the two circles (see *Figure 1983/2.1*).

The enlargement with centre K maps C_1 to C_2, O_1 to O_2, P_1 and Q_1 to P_2 and Q_2, respectively. Hence the image of M_1 is M_2; let A' denote the image of A.

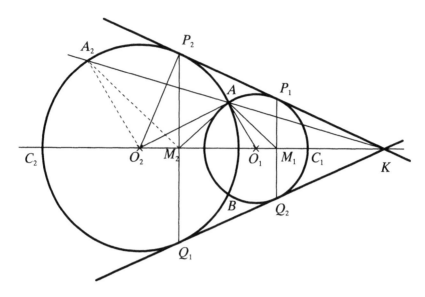

Figure 83/2.1

The enlargement preserves the angles, so $\angle O_1 A M_1 = \angle O_2 A_2 M_2$, hence it is enough to prove that

(2) $$\angle O_2 A M_2 = \angle O_2 A_2 M_2.$$

Observe that the power of K with respect to the circle C_2 is $KP_2^2 = KA \cdot KA_2$. Applying it to the right triangle KP_2O_2 we get $KP_2^2 = KM_2 \cdot KO_2$, hence

$$KA \cdot KA_2 = KM_2 \cdot KO_2,$$

therefore the points $AA_2O_2M_2$ lie on the same circle, thus A and A_2 are the angles of circumference of the chord M_2O_2, consequently (2) and so (1) holds. We proved the statement.

Second solution. As in the first solution we shall prove that the equality (1) holds. Let F denote the point of intersection of P_1P_2 and h, the common chord of C_1 and C_2. F is the midpoint of P_1P_2, because its power with respect to the two circles is the same, hence h is the median of the symmetric trapezium $P_1Q_1Q_2P_2$, and so it is the perpendicular bisector of the segment M_1M_2 (see *Figure 1983/2.2*).

Let O_1' denote the reflection of O_1 to h, then $\angle O_1 A M_1 = \angle O_1' A M_2$. In order to prove (1) we have to show that $\angle O_1' A M_2 = \angle O_2 A M_2$, that is AM_2 is a bisector of the triangle $O_1' A O_2$. For this, it is enough to prove that

(3) $$\frac{AO_2}{AO_1'} = \frac{M_2O_2}{O_1'M_2}.$$

The ratio of the enlargement mapping C_1 to C_2 is the ratio of their radii $\frac{AO_2}{AO_1}$. Because of the reflection, $AO_1 = AO_1'$, hence the ratio is $\frac{AO_2}{AO_1'}$. As the

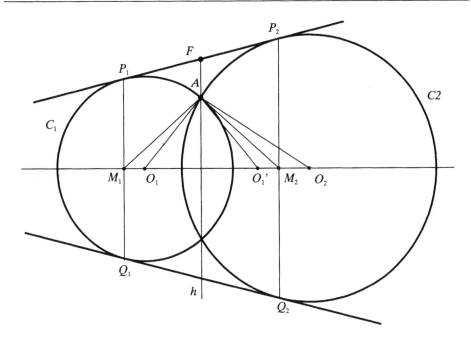

Figure 83/2.2

enlargement maps M_1O_1 to M_2O_2 and by the symmetry $M_1O_1 = O'_1M_2$,

$$\frac{AO_2}{AO'_1} = \frac{M_2O_2}{M_1O_1} = \frac{M_2O_2}{O'_1M_2},$$

hence we proved (3), and so the statement of the theorem.

1983/3. *Let a, b, c denote pairwise coprime positive integers. Prove that*
$$2abc - ab - bc - ca$$
is the largest integer which cannot be expressed as

(1) $$xbc + yca + zab,$$

where x, y, z are non negative integers.

First solution. Let $S = 2abc - ab - bc - ca$. First we show that S cannot be written in the desired form. Indeed,
$$2abc - ab - bc - ca = xbc + yca + zab,$$

implies

(2) $$2abc = (x+1)bc + (y+1)ca + (z+1)ab.$$

As a divides the l.h.s, it divides the r.h.s. so it has to divide $x+1$, hence there is a positive integer a' such that $x+1 = aa'$ holds. Similarly, there are positive integers b' and c' such that
$$y+1 = bb', \qquad z+1 = cc'.$$
Substituting these values to (2) and dividing by abc we get that
$$2 = a' + b' + c'.$$
That is impossible as 2 is not the sum of three positive integers.

Now, we have to show that for an arbitrary positive integer k there are positive integers x, y, z such that:

$$2abc + k = xbc + yca + zab.$$

Consider the triples (x, y, z), where

(3) $$1 \le x \le a, \quad 1 \le y \le b, \quad 1 \le z \le c.$$

In the set H of triples under (1) with these coefficients there are abc many elements. The remainder of any two elements of H mod abc is different. Indeed, if the remainder using the triples (x_1, y_1, z_1) and (x_2, y_2, z_2) as coefficients agree, consider the difference of the two numbers:

$$(x_1 - x_2)bc + (y_1 - y_2)ca + (z_1 - z_2)ab.$$

As a, b, c are coprime, the difference is divisible by a, b and c. Now, by (3) it can only hold if $x_1 = x_2$, $y_1 = y_2$, $z_2 = z_2$.

Thus all mod abc remainders occur in H. Now, let (x, y, z) be a triple satisfying conditions (3), such that the expression in (1) gives the same residue as k. abc divides $k - (xbc + yca + zab)$ and so $k - (xbc + yca + zba) + 2abc$ is divisible by abc, too, thus there is a positive integer n such that

$$k - (xbc + yca + zab) + 2abc = nabc,$$

that is

(4) $$2abc + k = nabc + xbc + yca + zab.$$

Using values x, y, z satisfying (3) we get:

$$xbc + yca + zab \le 3abc,$$

hence (4) implies

$$k \le nabc - 2abc + 3abc = (n+1)abc,$$

as $k > 0$ and $n \ge 0$. So (4) can be rewritten as:

$$2abc + k = (na + x)bc + yca + zab,$$

hence with the positive integers $na + x = x'$, y, z we presented $2abc + k$ in the desired form.

Second solution. $S = 2abc - ab - bc - ca$ is not of the form $xbc + yca + zab$. Indeed, in case

$$xbc + yca + zab = 2abc - ab - bc - ca,$$

(5) $$(x+1)bc + (y+1)ca + (z+1)ab = 2abc$$

holds. As, for example, bc and a are coprime, a divides $x+1$ hence $a \le x+1$; similarly, $b \le y+1$ and $c \le z+1$, implying

$$2abc = (x+1)bc + (y+1)ca + (z+1)ab \ge 3abc$$

that is impossible.

Now, we prove that for an arbitrary positive integer k' there are x, y, z positive integers such that

$$(6) \qquad xbc + yca + zab = 2abc + k'$$

holds. Or, equivalently, if k is a positive integer such that $k > 2abc$, then it is of the form

$$(7) \qquad xbc + yca + zab = k$$

where x, y, z are positive integers.

Observe that the numbers

$$bc, \ 2bc, \ 3bc, \ \ldots, \ (a-1)bc, \ abc$$

give distinct residues mod a since there are no two numbers among them with their difference divisible by a. Hence they form a complete residue system, and one of them, e.g. $x_1 bc$ gives the same residue mod a as k. Thus

$$x_1 bc \equiv k \quad (\text{mod } a), \qquad 1 \leq x_1 \leq a.$$

Similarly, we get the positive integers y_1, z_1 such that

$$y_1 ca \equiv k \quad (\text{mod } b) \qquad 1 \leq y_1 \leq b,$$
$$z_1 ab \equiv k \quad (\text{mod } c) \qquad 1 \leq z_1 \leq b.$$

This implies that e.g.

$$(x_1 bc - k) + y_1 ca + z_1 ab = x_1 bc + y_1 ca + z_1 ab - k$$

is divisible by a, and similarly, by b and c, too; moreover, as these numbers are coprime, it is divisible by abc, thus

$$s = x_1 bc + y_1 ca + z_1 ab \equiv k \quad (\text{mod } abc).$$

This also means that s and k (thus $s-1$ and $k-1$) give the same residue mod abc, that is

$$(8) \qquad k - 1 = q \cdot abc + r,$$
$$(9) \qquad s - 1 = q' \cdot abc + r \qquad (0 \leq r < abc),$$

where, $k > 2abc$ implies $q \geq 2$, and the assumptions for x_1, y_1, z_1 imply $s \leq 3abc$, and so $q' \leq 2$.

Considering the difference of (8) and (9) we obtain

$$k - s = (q - q')abc, \qquad (q - q' \geq 0)$$
$$k = s + (q - q')abc = (x_1 + (q - q')a)bc + y_1 ca + z_1 ab,$$

hence choosing $x = x_1 + (q - q')a$, $y = y_1$, $z = z_1$ we get (7) and this is what we wanted.

Remark. Similarly, it can be proved that if the positive integers $a_1, a_2, \ldots,$ a_n are pairwise coprime and $s = a_1 a_2 \ldots a_n$, then the largest number that is not of the form

$$x_1\frac{s}{a_1} + x_2\frac{s}{a_2} + \ldots + x_n\frac{s}{a_n}$$

with $x_1, x_2, \ldots, x_k$ integers is

$$(n-1)s - \left(\frac{s}{a_1} + \frac{s}{a_2} + \ldots + \frac{s}{a_n}\right).$$

1983/4. *Let ABC be an equilateral triangle and E the set of all points contained in the three segments AB, BC and CA (including A, B and C). Determine whether, for every partition of E into two disjoint subsets, at least one of the two subsets contains the vertices of a right-angled triangle.*

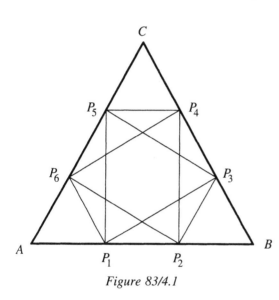

Figure 83/4.1

Solution. We show that the answer to the statement is affirmative. Let P_1, P_2, P_3, P_4, P_5, P_6 denote the points dividing the sides AB, BC, CA into three equal parts, respectively. These points are the vertices of a regular hexagon. We shall use only these vertices along with the vertices of the triangle. We show that these nine points cannot be divided into two parts avoiding a monochromatic right triangle. (see *Figure 1983/4.1*).

For simplicity, let us colour the points of the two sets by green and red. As two of the points A, B, C have the same colour, we may assume that A and B are red. Considering all possible colourings of the points P_1 and P_2 on the side AB we show that the assumption that there is no right triangle leads to a contradiction.

1. P_1 and P_2 are red. P_3 and P_6 has to be green, as the triangles P_1P_2B or P_2P_6A were red. P_5 is green, because of the triangle AP_1P_5, but then the triangle $P_3P_5P_6$ is green, contradiction.

2. P_1 and P_2 are green. P_4 and P_5 has to be red because of the triangles $P_2P_1P_5$, or $P_1P_2P_4$; P_6 and P_3 are green for the triangles BP_4P_6 and AP_5P_3; but, now, $P_3P_5P_6$ is red, contradiction.

3. P_1 red, P_2 green (or vice verse). Now, P_3 and P_5 are green because of the triangles P_1P_3B and AP_1P_5. Now, $P_2P_3P_5$ is green, contradiction again.

Thus we proved that there is a monochromatic triangle.

1983/5. *Is it possible to choose 1983 distinct positive integers, all less than or equal to 10^5, no three of which are consecutive terms of an arithmetic progression?*

First solution. The answer is affirmative. Let us choose the set consisting of all positive integers whose base 3 representations contain only the digits 0 and 1 (i.e. no 2-s). Since $3^{10} < 10^5 < 3^{11}$, these numbers have at most 11 digits; the largest one among them is $11\,111\,111\,111_3 = 88\,573_{10}$. The number of these numbers is $2^{11} - 1 = 2047$ (0 is not among them).

We show that these 2047 numbers do not contain an arithmetic progression. If x, y and z are three elements of an arithmetic progression, then $x + z = 2y$. The number $2y$ contains only digits 0 and 2. Hence x and z must match digit by digit thus $x = z$, which is impossible. Hence the chosen 2047 numbers contain no arithmetic progression of length 3, therefore the selection is possible.

Second solution. Let us define a sequence of sets on the following way: Let $H_1 = \{1,\ 2\}$ and let H_{n+1} be defined such that first construct the set S_n, in a way that we add 3^n to every element of H_n (e.g. $S_2 = \{4,\ 5\}$), and let

(1) $$H_{n+1} = H_n \cup S_n.$$

Thus:

$$H_2 = \{1,\ 2;\ 4,\ 5\}, \qquad H_3 = \{1,\ 2,\ 4,\ 5;\ 10,\ 11,\ 13,\ 14\}.$$

Then the size of H_n is 2^n. We prove by induction that the largest element of H_n is $\dfrac{3^n + 1}{2}$. The largest element of H_1 is $\dfrac{3+1}{2} = 2$; suppose that the largest element of H_{n-1} is $\dfrac{3^{n-1} + 1}{2}$, hence the largest element of H_n is

$$\frac{3^{n-1} + 1}{2} + 3^{n-1} = \frac{3^n + 1}{2}.$$

Thus the elements of H_n are in the interval $\left[1, \dfrac{3^n + 1}{2}\right]$ and the elements of S_n in the interval $\left[3^n + 1, \dfrac{3^{n+1} + 1}{2}\right]$.

We prove by induction that the sets H_i contain no arithmetic progressions of length three. For H_1 and H_2 this is obviously true, and suppose that it holds for H_n, too. Consider the set $H_{n+1} = H_n \cup S_n$; we show that if in this set $x < y < z$, then $x + z = 2y$ is impossible. This is the condition of being three consecutive members of an arithmetic progression.

If x and z are from H_n, then the statement holds by the assumption. If x and z are from S_n, then there are some x' and z' in H_n such that

$$x = x' + 3^n, \qquad z = z' + 3^n$$

and so

$$y = \frac{x+z}{2} = \frac{x'+z'}{2} + 3^n.$$

This is not in H_n as it is greater than the largest element of H_n, but it is not in S_n, either, because then $\dfrac{x''+z'}{2}$ were in H_n, contradicting our assumption, thus y is not in H_{n+1}.

Finally, if $x \in H_n$ and $z \in S_n$, then $x \geq 1$, $z \geq 3^n + 1$, and so

$$y = \frac{x+z}{2} \geq \frac{3^n + 1 + 1}{2} > \frac{3^n + 1}{2},$$

thus it is greater than the largest element of H_n, and

$$y = \frac{x+z}{2} \leq \frac{1}{2}\left(\frac{3^n+1}{2} + \frac{3^{n+1}+1}{2}\right) = \frac{1}{2}(2 \cdot 3^n + 1) < 3^n + 1,$$

thus it is smaller than the smallest element of S_n, hence it is not in H_{n+1}. Therefore H_{n+1} contains no arithmetic progression of length 3.

We proved in general that there are 2^n numbers not greater than $\dfrac{3^n+1}{2}$ such that they contain no arithmetic progression of length 3.

For $n = 11$ the result says that there are 2048 numbers among the first 88 574 integers not containing an arithmetic progression of length 3, so the answer to the question is positive.

1983/6. *Let a, b, c be the length of the sides of a triangle. Prove that*

(1) $$a^2 b(a-b) + b^2 c(b-c) + c^2 a(c-a) \geq 0.$$

First solution. Introduce the following notations:

$$x = -a+b+c, \qquad y = a-b+c, \qquad z = a+b-c,$$

thus, x, y, z are twice the length of the segments between the vertices and the touching points of the incircle. So

$$a = \frac{y+z}{2}, \qquad b = \frac{z+x}{2}, \qquad c = \frac{x+y}{2}.$$

Substitute them to (1) and multiply the inequality by 16:

$$(y+z)^2(z+x)(y-x) + (z+x)^2(x+y)(z-y) + (x+y)^2(y+z)(x-z) \geq 0.$$

After reordering it reduces to:

(2) $$x^3 z + y^3 x + z^2 y \geq x^2 yz + y^2 zx + z^2 xy,$$

and so

$$x^3 z + y^3 x + z^3 y - x^2 yz - y^2 zx - z^2 xy =$$

$$= zx(x-y)^2 + xy(y-z)^2 + yz(z-x)^2 \geq 0.$$

The validity of the inequality is clear, as all the summands are nonnegative.

As $x > 0$, $y > 0$, $z > 0$, equality holds if and only if $x = y = z$, that is if $a = b = c$, thus the triangle is equilateral.

Second solution. Transforming both sides to polynomials it is easy to see that:

$$a^2b(a-b) + b^2c(b-c) + c^2a(c-a) = a(b+c-a)(b-c)^2 + b(a+b-c)(a-b)(a-c).$$

Since the l.h.s. of (1) remains the same at the permutation $a \to b$, $b \to c$, $c \to a$, we may assume that $a \geq b$, c; now, on the r.h.s. of (1) every term is positive thus we proved the statement.

Remarks. 1. (1) holds in some cases when the numbers are not the sides of a triangle, e.g. if $a = 1$, $b = 3$, $c = 5$.

2. (2) can be shown in different ways. e.g. dividing both sides of (2) by xyz we obtain:

(3) $$\frac{x^2}{y} + \frac{y^2}{z} + \frac{z^2}{x} \geq x + y + z.$$

Now, the application of the Cauchy inequality for the triples $\left(\dfrac{x}{\sqrt{y}}, \dfrac{y}{\sqrt{z}}, \dfrac{z}{\sqrt{x}} \right)$ and $(\sqrt{x}, \sqrt{y}, \sqrt{z})$ gives:

$$(x+y+z)^2 = \left(\frac{x}{\sqrt{y}}\sqrt{y} + \frac{y}{\sqrt{z}}\sqrt{z} + \frac{z}{\sqrt{x}}\sqrt{x} \right)^2 \leq \left(\frac{x^2}{y} + \frac{y^2}{z} + \frac{z^2}{x} \right)(x+y+z),$$

and dividing by $(x+y+z)$ we arrive to (3).

1984.

1984/1. *Prove that*

(1) $$0 \leq xy + yz + zx - 2xyz \leq \frac{7}{27},$$

where x, y and z are nonnegative real numbers for which

(2) $$x + y + z = 1.$$

First solution. To prove the left side of the inequality, observe that (2) implies $0 \leq x, y, z \leq 1$ and so

$$xy + yz + zx - 2xyz = xy(1-z) + yz(1-x) + zx \geq 0,$$

since all three terms of the l.h.s. are nonnegative.

In order to show the right side of the inequality, introduce the notations

$$x = a + \frac{1}{3}, \qquad y = b + \frac{1}{3}, \qquad z = c + \frac{1}{3}$$

now, (2) gives

(3) $$\qquad\qquad a + b + c = 0, \qquad -\frac{1}{3} \le a,\, b,\, c \le \frac{2}{3}.$$

After substituting $b + c = -a$, we obtain

(4)

$$xy + yz + zx - 2xyz = \frac{7}{27} + \frac{1}{3}(ab + bc + ca - 6abc) = \frac{7}{27} + \frac{1}{3}\left(bc - a^2 - 6abc\right).$$

Our original expression is symmetric in a, b, c, hence we may assume that $a \le b \le c$. According to (3), a, b, c cannot be all positive or negative at the same time, so there are two possibilities:

$\alpha)$ $a \le b \le 0 \le c$,   or $\beta)$ $-\frac{1}{3} \le a \le 0 \le b \le c$.

In order to prove (2) we have to show that the expression in the parenthesis in (4) is not positive. In case $\alpha)$ every expression in the parenthesis is positive, thus (2) holds.

In case $\beta)$, we rearrange the expression for our purposes:

$$bc - a^2 - 6abc = bc - (b+c)^2 - 6abc = -(b-c)^2 - 3bc(1+2a).$$

Now, as $1 + 2a > 0$ and $bc \ge 0$, the expression cannot be positive, thus (2) holds.

Second solution. (2) implies that at least one of x, y, z, let us say x, is not greater than $\frac{1}{3}$, thus $1 - 2x > 0$, and so

$$xy + yz + zx - 2xyz = xy + yz(1 - 2x) + zx \ge 0,$$

hence we proved the left hand side of (1).

To prove the right hand side, we distinguish two cases:

$\alpha)$ one of x, y, z, let us say z, is at least $\frac{1}{2}$; $z \ge \frac{1}{2}$, $\beta)$ $0 \le x,\, y,\, z \le \frac{1}{2}$.

In case $\alpha)$, using $x + y = 1 - z$ we get

$$xy + yz + zx - 2xyz = z(x+y) + xy(1 - 2z) =$$

$$= z(1-z) + xy(1-2z) \le z(1-z) \le \frac{1}{4} < \frac{7}{27}.$$

In case $\beta)$ apply the following substitutions:

$$a = 1 - 2x, \qquad b = 1 - 2y, \qquad x = 1 - 2z.$$

These values give

(5) $$\qquad\qquad a,\, b,\, c \ge 0, \qquad a + b + c = 1.$$

Substituting and applying (5) we obtain:

(6)
$$xy + yz + zx - 2xyz = \frac{1+abc}{4}.$$

Now, the A.M.–G.M. inequality gives

$$abc \le \left(\frac{a+b+c}{3}\right)^3 = \frac{1}{27},$$

hence (6) implies

$$xy + yz + zx - 2xyz \le \frac{\frac{28}{27}}{4} = \frac{7}{27},$$

and so we proved (1) in every case.

Third solution. The problem involves the so called elementary symmetric polynomials of x, y, z that leads us to the following idea: consider the polynomial $f(t)$ of degree 3:

(7) $\quad f(t) = (t-x)(t-y)(t-z) = t^3 - t^2(x+y+z) + t(xy+yz+zx) - xyz.$

From (2) we get

$$2f\left(\frac{1}{2}\right) = \frac{1}{4} - \frac{1}{2} + (xy+yz+zx) - 2xyz,$$

$$xy + yz + zx - 2xyz = 2f\left(\frac{1}{2}\right) + \frac{1}{4}.$$

The problem says that

$$0 \le 2f\left(\frac{1}{2}\right) + \frac{1}{4} \le \frac{7}{27},$$

that is

(8)
$$-\frac{1}{8} \le f\left(\frac{1}{2}\right) \le \frac{1}{216}.$$

We may assume that $x \ge y \ge z \ge 0$. Choose x, y, z such that $x \ge \frac{1}{2}$, $y \le \frac{1}{2}$, $z \le \frac{1}{2}$ holds. Then, (7) implies that $f\left(\frac{1}{2}\right) \le 0$, hence

$$-f\left(\frac{1}{2}\right) = \left(x - \frac{1}{2}\right)\left(\frac{1}{2} - y\right)\left(\frac{1}{2} - z\right).$$

Applying the A.M.–G.M. inequality and using that $x \le 1$

$$-f\left(\frac{1}{2}\right) \le \left(\frac{x-y-z+\frac{1}{2}}{3}\right)^3 = \left(\frac{2x-\frac{1}{2}}{3}\right)^3 \le \frac{1}{8}, \quad f\left(\frac{1}{2}\right) \ge -\frac{1}{8}.$$

If we choose the values of the variables such that $x \le \frac{1}{2}$ and so $y \le \frac{1}{2}$, $z \le \frac{1}{2}$ holds, the previous method gives:

$$0 \le f\left(\frac{1}{2}\right) \le \left(\frac{1}{2}-x\right)\left(\frac{1}{2}-y\right)\left(\frac{1}{2}-z\right) \le \left(\frac{\frac{3}{2}-(x+y+z)}{3}\right)^3 = \frac{1}{216}.$$

Thus we proved (8) and the statement of the theorem.

1984/2. *Find one pair of positive integers, a, b, such that:*

(1) $ab(a+b)$ *is not divisible by 7;*

(2) $(a+b)^7 - a^7 - b^7$ *is divisible by 7^7.*

First solution. Using the binomial theorem and observing that both $(a+b)^7$ and $a^7 + b^7$ is divisible by $(a+b)$, (2) reads as:

(3) $$(a+b)^7 - a^7 - b^7 = 7ab(a+b)(a^2+ab+b^2)^2.$$

Since $ab(a+b)$ is not divisible by 7, we must choose a and b such that 7^6 divides $\left(a^2+ab+b^2\right)^2$ or equivalently: a^2+ab+b^2 is divisible by $7^3 = 343$.

We try to choose a pair a, b such that $b = 1$, that is, for some integer k

(4) $a^2 + a + 1 = 343k,$ that is, $a^2 + a + (1 - 343k) = 0$

holds. This happens exactly if the discriminant of this equation of degree 2 in a is a perfect square, which means that

$$1 - 4(1 - 343k) = 1372k - 3$$

is a perfect square. This is true for $k = 1$, as $1369 = 37^2$. Hence (4) becomes:

$$a^2 + a - 342 = 0,   a = 18,   b = 1,$$

and the pair $(a, b) = (18, 1)$ satisfies the conditions, because 7 does not divide $18 \cdot 1 \cdot 19$.

Second solution (due to Géza Kós). We finish the previous solution in a different way to determine all pairs a, b that satisfy the conditions of the problem; as we saw before, we need to find those pairs, a, b where 343 divides $a^2 + ab + b^2$, but a, b, $a+b$ are not divisible by 7. We start with the identity

$$18\left(a^2+ab+b^2\right) = 343ab + (a - 18b)(18a - b).$$

18 and 343 are coprime so only one of $a - 18b$ and $18a - b$ can be divisible by 7. Otherwise their difference, $17(a+b)$ is divisible by 7; hence there are two cases:

A) $a = 18b + 343k,$

where 7 does not divide b and k is an integer such that $18b + 343k$ is positive;

B) $b = 18a + 343k,$

where 7 does not divide a and k is an integer such that $18a + 343k$ is positive;

It is clear that in none of the two cases is $a + b$ divisible by 7. Indeed,

$$a + b = 19b + 343k, \qquad \text{and} \qquad a + b = 19a + 343k.$$

The values given in A) and B) provide all solutions to the problem.

1984/3. *In the plane two different points O and A are given. For each point $X \neq O$ on the plane denote by $\omega(X)$ the measure of the angle between OA and OX in radians counterclockwise from OA $(0 \leq \omega(X) < 2\pi)$. Let $C(X)$ be the circle with centre O and radius $OX + \dfrac{\omega(X)}{OX}$. Each point of the plane is coloured by one of a finite number of colours. Prove that there exists a point Y for which $\omega(Y) > 0$ such that its colour appears on the circumference of the circle $C(Y)$.*

Solution. Consider the set of all concentric circles with centre O whose radii are less than 1. Assign to every circle the set of colours that appear on the circle. As there are finitely many colours, we have two circles with the same combination of colours. Call them B and S with radii b and s respectively, and let $b < s < 1$ (see *Figure 1984/3.1*).

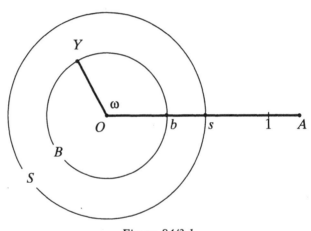

Figure 84/3.1

Pick a point Y on B such that $\omega(Y) = b(s - b)$ holds. Such a point, Y, exists, because $0 < b(s - b) < 1 < 2\pi$. Hence

$$s = b + \frac{\omega(Y)}{b},$$

the circle S is the same as $C(Y)$, and as S and B bear the same set of colours, $C(Y) = S$ contains the colour of Y, and this is what we wanted to prove.

1984/4. *Let $ABCD$ be a convex quadrilateral with the line CD tangent to the circle on diameter AB. Prove that the line AB is tangent to the circle on diameter CD if and only if BC and AD are parallel.*

Solution. Let F and G denote the midpoints of the sides AB and CD; F' denotes the foot of the perpendicular from F to CD, G' the one from G to AB

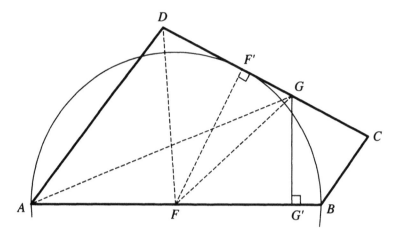

Figure 84/4.1

(see *Figure 1984/4.1*). From the conditions $AF = FB = FF'$, and so

$$\frac{AF}{FF'} = 1.$$

The radius of the circle over CD is $CG = GD$ that touches the AB side if and only if $GG' = GD$, that is

$$\frac{GD}{GG'} = 1.$$

It can be restated as: the circle with diameter CD touches AB if and only if:

(1) $$\frac{AF}{FF'} = \frac{GD}{GG'}, \quad \text{that is,} \quad \frac{AF \cdot GG'}{2} = \frac{GD \cdot FF'}{2}.$$

This means that the area of the triangles AFG and DFG agrees. As FG is a common side of the two triangles, the equality of their areas and the convexity of $ABCD$ imply that A and D are of the same distance from FG (on the same side of FG), thus AD and FG are parallel. FG is the median of the quadrangle $ABCD$, and the median is parallel to a side if and only if the quadrangle is a trapezium, that is if AD and BC are parallel; thus we proved our statement.

1984/5. *Let d be the sum of the lengths of all the diagonals of a plane convex polygon with $n \geq 3$ vertices. Let p be its perimeter. Prove that:*

(1) $$n - 3 < \frac{2d}{p} < \left[\frac{n}{2}\right] \left[\frac{n+1}{2}\right] - 2.$$

Solution. First we prove the lower bound that is equivalent to the inequality:

$$2d > (n - 3)p.$$

Choose two non-adjacent sides of the polygon XY and ZU such that the quadrangle $XYZU$ is convex (see *Figure 1984/5.1*). The triangle inequality implies that the sum of the two diagonals XZ and YU is greater than the sum of the two opposite sides XY and ZU.

(2) $$XZ + YU > XY + ZU.$$

A diagonal XZ is the diagonal of two quadrangles, because a pair of opposite sides of a quadrangle can be chosen in two ways on the two sides of XZ. So, if we choose all possible quadrangles on the way described above, and sum the inequalities of type (2), the l.h.s. is $2d$. For a side XY we can choose an opposite side on $n - 3$ many ways, so for the XY side we can choose $n - 3$ quadrangles. But we choose a quadrangle n times when we make all possible choices. Hence the number of distinct quadrangles is $n - 3$ and the perimeter of the polygon comes up twice among their sides. Hence we get

$$2d > (n - 3)p,$$

and we proved the left side of (1).

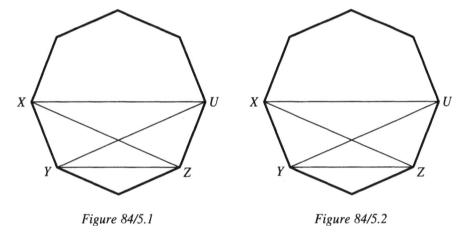

Figure 84/5.1   Figure 84/5.2

In order to prove the right side, let first choose $n = 2k + 1$. The diagonal is the shortest path between its endpoints, it is shorter than each polygonial path connecting them. We shall use the polygonial path that uses less terms. For example we estimate the $A_1 A_2 \ldots A_n$ part of the polygonial path with the diagonals from A_i in the following way: (see *Figure 1984/5.2*) ($A_{n+1} \equiv A_1$):

$$A_i A_{i+2} < A_i A_{i+1} + A_{i+1} A_{i+2}$$
$$A_i A_{i+3} < A_i A_{i+1} + A_{i+1} A_{i+2} + A_{i+2} A_{i+3}$$

(3) $\vdots$

$$A_i A_{i+k-1} < A_i A_{i+1} + \ldots + A_{i+k-2} A_{i+k-1}$$
$$A_i A_{i+k} < A_i A_{i+1} + \ldots + A_{i+k-1} A_{i+k}$$

Proceed the same way for every diagonal and sum these inequalities. On the l.h.s every diagonal occurs twice, hence the sum is d. On the r.h.s.

$$2 + 3 + \ldots + k = \frac{1}{2}(k - 1)(k + 2) = \frac{1}{2}(k(k + 1) - 2)$$

many sides come up at a vertex, so there are $\dfrac{n}{2}(k(k+1)-2)$ sides in the sum. None of the sides were distinguished, and since each side occurs equally, on the r.h.s. we have

$$\frac{p}{2}(k(k+1)-2),$$

that is

(4)
$$d < \frac{p}{2}(k(k+1)-2)).$$

As $\dfrac{n}{2}=k+\dfrac{1}{2}$, $\left[\dfrac{n}{2}\right]=k$, moreover $\dfrac{n+1}{2}=k+1$ and so $\left[\dfrac{n+1}{2}\right]=k+1$, (4) rewrites as:

$$\frac{2d}{p} < \left[\frac{n}{2}\right]\left[\frac{n+1}{2}\right]-2.$$

Thus we proved (1) for odd many vertices.

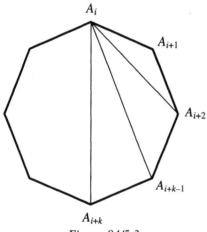

A_i

A_{i+1}

A_{i+2}

A_{i+k-1}

A_{i+k}

Figure 84/5.3

For $n=2k$, in (3) we omit the last row, and substitute it by the obvious estimate for the diagonal $A_i A_{i+k}$ (see *Figure 1984/5.3*):

$$A_i A_{i+k} < \frac{p}{2} \qquad (i=1, 2, \ldots, k).$$

At the summation the r.h.s. becomes the sum d of the diagonals. On the r.h.s a vertex gives

$$2+3+\ldots+k-1=\frac{1}{2}(k-2)(k+1)$$

many sides and repeating the previous arguments this is $\dfrac{1}{2}(k-2)(k+1)$ times the perimeter of the polygon; considering the $\dfrac{kp}{2}$ sum coming from the estimates of the diagonals $A_i A_{i+k}$ we get

$$d < \frac{p}{2}(k+(k-2)(k+1))=\frac{p}{2}\left(k^2-2\right).$$

For $n=2k$, $k=\left[\dfrac{n}{2}\right]=\left[\dfrac{n+1}{2}\right]$ and so $k^2=\left[\dfrac{n}{2}\right]\left[\dfrac{n+1}{2}\right]$, therefore our inequality becomes

$$\frac{2d}{p} < \left[\frac{n}{2}\right]\left[\frac{n+1}{2}\right]-2,$$

hence (1) holds for every $n>3$.

1984/6. *Let a, b, c, d be odd numbers, such that:*

(1) $$0 < a < b < c < d,$$

(2) $$ad = bc,$$

(3) $$a + d = 2^k, \qquad b + c = 2^m$$

holds, where k and m are integers. Prove that $a = 1$.

Solution. In our solution we construct the integers a, b, c, d as functions of k and m. First we show that $k > m$. Indeed, as $c = \dfrac{ad}{b}$ we have

$$2^k - 2^m = (a + d) - (b + c) = (b - a)\left(\frac{d}{b} - 1\right) > 0.$$

Now substitute d and c expressed from (3) into (2):

$$a\left(2^k - a\right) = b(2^m - b),$$

$$2^m b - 2^k a = b^2 - a^2 = (b + a)(b - a).$$

Hence 2^m divides $(b + a)(b - a)$. $b + a$ and $b - a$ are even but they cannot be both divisible by 4, since their sum $(2b)$, is divisible only by the first power of 2. Hence one of $a - b$ and $a + b$ is divisible by 2 and the other one by 2^{m-1}. As

$$b - a < b < \frac{b + c}{2} = 2^{m-1}$$

and

(4) $$b + a < b + c = 2^m,$$

only the first power of 2 divides $b - a$ and so $b + a$ is divisible by 2^{m-1}. $b + a = {} = r \cdot 2^{m-1}$ (r is a positive integer), but (4) implies $r < 2$, thus $r = 1$, hence

(5) $$b + a = 2^{m-1}.$$

Moreover:

(6) $$c - a = (b + c) - (b + a) = 2^m - 2^{m-1} = 2^{m-1}.$$

a and b, a and c are coprime because they are odd numbers, and any common divisor of them divides 2^{m-1} by (5) and (6). On the other hand (2) implies that a divides bc, thus

(7) $$a = 1.$$

With this we proved the statement of the problem. (5), (6), (2) and (7) gives

(8) $$b = 2^{m-1} - 1, \qquad c = 2^{m-1} + 1, \qquad d = bc = 2^{2(m-1)} - 1,$$

combining with (3) we get $k = 2(m - 1)$. It is easy to see that the numbers in (8) are all the numbers that satisfy the conditions for $m > 2$.

1985.

1985/1. *A circle has centre on the side AB of the cyclic quadrilateral ABCD. The other three sides are tangents to the circle. Prove that*

$$AD + BC = AB.$$

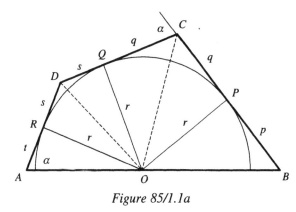

Figure 85/1.1a

First solution. Let O denote the centre of the touching circle, r its radius and P, Q, R the points of tangency of the sides BC, CD and DA respectively The length of the segments are denoted by $BP = p$, $PC = CQ = q$, $QD = DR = s$, $RA = = t$ (see *Figure 1985/1.1a*). With our notations we have to show that

$$AB = p + q + t + s.$$

As $ABCD$ is cyclic, $\angle BAD = \alpha$ equals the external angle at C.

Cut and change the right triangles AOR, BOP and fix the other parts of the quadrangle. We get the quadrangle $A'CDB'$ that touches the circle with radius r and centre O at three sides.

$A'CDB'$ is a trapezium, because the sum of the angles at A' and C is $180°$ (see *Figure 1985/1.1b*).

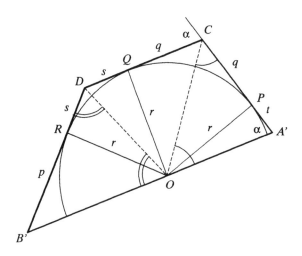

Figure 85/1.1b

$PCQO$ is a deltoid, so OC is a bisector, hence $\angle OCA' = 90° - \dfrac{\alpha}{2}$. This implies that $\angle A'OC = 180° - \alpha - \left(90° - \dfrac{\alpha}{2}\right) = 90° - \dfrac{\alpha}{2}$, hence $A'OC$ is an isosceles triangle, thus $OA' = t + q$. Similarly, $OB' = p + s$, thus

$$AB = A'B' = A'O + OB' = t + q + p + s,$$

and this is what we wanted to prove.

Second solution. Let O be the centre of the circle, k the circumcircle of the triangle CDO that intersects the AB side in a second point P. P is an

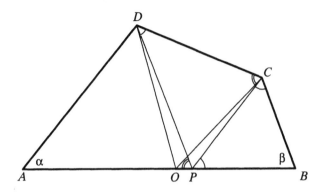

Figure 85/1.2

inner point of AB; if $P = O$ then AB touches the circle CDO at O (see *Figure 1985/1.2*). As the circle with centre O touches the sides AD, DC, BC the lines OD and OC bisect the angles at D and C. Set $\angle A = \alpha$, $\angle B = \beta$.

Assume that P lies on the ray OB. As the quadrangle $OPCD$ is cyclic, $\angle BPC = \angle ODC = 90° - \dfrac{\beta}{2}$, hence in the triangle PBC we have $\angle PCB = 180° - \beta - \left(90° - \dfrac{\beta}{2}\right) = 90° - \dfrac{\beta}{2}$, thus PCB is an isosceles triangle, therefore $PB = BC$ (if $O = P$, $\angle BPC = \angle ODC = 90° - \dfrac{\beta}{2}$ holds, as well).

As they are measured by the same arc in k,

$$\angle APD = \angle OPD = \angle OCD = 90° - \dfrac{\alpha}{2},$$

hence $\angle ADP = 180° - \alpha - \left(90° - \dfrac{\alpha}{2}\right) = 90° - \dfrac{\alpha}{2}$, and so PAD is an isosceles triangle, implying $AP = AD$. Combining these two results:

$$AB = AP + PB = AD + BC.$$

Third solution. Use the notations of *Figure 1985/1.1a*. OC and OD bisect the angles of the quadrangle at C and D, so $\angle QCP = 180° - \alpha$, hence $\angle OCP =$ $= 90° - \dfrac{\alpha}{2}$ and in the right triangle OPC $q = r \cot\left(90° - \dfrac{\alpha}{2}\right) = r \tan \dfrac{\alpha}{2}$. From the right triangle ARO, $t = r \cot\alpha$, $AO = \dfrac{r}{\sin\alpha}$; hence

$$t + q = r\left(\tan\frac{\alpha}{2} + \cot\alpha\right).$$

As

$$\tan\frac{\alpha}{2} + \cot\alpha = \frac{\sin\frac{\alpha}{2}}{\cos\frac{\alpha}{2}} + \frac{\cos\alpha}{\sin\alpha} = \frac{2\sin^2\frac{\alpha}{2} + 1 - 2\sin^2\frac{\alpha}{2}}{2\sin\frac{\alpha}{2}\cos\frac{\alpha}{2}} = \frac{1}{\sin\alpha},$$

$$t + q = \frac{r}{\sin\alpha} = AO.$$

Changing the roles of PC and RD, AR and BP, the previous arguments give

$$p + s = OB.$$

Adding these results we get what we wanted.

1985/2. *Let n and k be relatively prime positive integers with $k < n$. Each number in the set $M = \{1, 2, 3, \dots, n-1\}$ is coloured either blue or white, such that:*

(a) *for each i in M, both i and $n - i$ have the same colour;*

(b) *for each i in M not equal to k, both i and $|i - k|$ have the same colour.*

Prove that all numbers in M must have the same colour.

Solution. Let $a \sim b$ denote if a and b have the same colour. We start with two lemmas:

α) if $a \equiv b \pmod{k}$, then $a \sim b$ [30];

β) there exists a complete residue system mod k, $b_1, b_2, \dots, b_k$, such that $b_i \sim b_{i+1}$ $(i = 1, 2, \dots, k-1)$.

These imply our statement: β) means that all the b_i-s have the same colour, but every element of M belongs to a residue class mod k, hence they have the same colour.

We assume that $k > 1$, because in case $k = 1$ (b) implies $i \sim |k - i| = i - 1$, thus any two consecutive members of M have the same colour, hence it holds for every element of M.

α) Let $a = sk + r$ $(0 < r \le k)$.

(b) implies that $r \sim r + k$, as $r + k \sim |k - (r+k)| = r$;

$$r \sim r + k \sim r + 2k \sim \dots \sim r + sk = a,$$

implying that the elements in the residue class of r mod k have the same colour.

β) Consider the numbers

$$n, \ 2n, \ 3n, \ \ldots, \ kn.$$

Their residues mod k are:

$$b_1, \ b_2, \ b_3, \ \ldots, \ b_k,$$

thus $1 \le b_i \le k$ ($i = 1, 2, \ldots, k$). Here, b_k is k. The b_i-s form a complete residue system mod k. Indeed, if b_i and b_j ($b_i > b_j$) were in the same class, their difference is

$$b_i - b_j \equiv (i - j)n \equiv 0 \pmod{k},$$

which is impossible as k and n are coprime and $i - j < k$.

Set $1 \le i \le k - 1$. (b) implies $b_i \sim |k - b_i|$, and as $k - b_i > 0$, $b_i \sim k - b_i$. (a) gives

$$k - b_i \sim n - (k - b_i) = n - k + b_i.$$

Now

$$n - k + b_i \equiv n + b_i \equiv n + in = n(i + 1) \equiv b_{i+1},$$

that is $b_i \sim n - k + b_i \sim b_{i+1}$, $b_i \sim b_{i+1}$, and we proved the statement of the problem.

1985/3. *For any polynomial $P(x) = a_0 + a_1 x + \ldots + a_k x^k$ with integer coefficients let $\omega(P)$ denote the number of odd coefficients, and let $Q_i(x) = (1 + x)^i$, where $i = 0, 1, 2, \ldots$.*

Prove that if $i_1, i_2, \ldots, i_n$ are integers such that $0 \le i_1 < i_2 < \ldots < i_n$, then

(1) $$\omega\left(Q_{i_1} + Q_{i_2} + \ldots + Q_{i_n}\right) \ge \omega\left(Q_{i_1}\right).$$

Solution. We start with two observations:

Lemma 1. If $0 < k < 2^m$ ($m \ge 1$ integer), then $\binom{2^m}{k}$ is even.

The definition of the binomial coefficients implies:

$$k \binom{2^m}{k} = 2^m \binom{2^{m-1} - 1}{k - 1}$$

as $k < 2^m$, it shows that $\binom{2^m}{k}$ is divisible by 2, hence it is even.

Lemma 2. If $P(x)$ is a polynomial of degree n, where $n < 2^m$ ($m \ge 1$ integer), then

(2) $$\omega(P \cdot Q_{2^m}) = 2\omega(P).$$

By the first Lemma

(3) $$Q_{2^m}(x) = (1 + x)^{2^m} = 1 + x^{2^m} + R(x),$$

where $R(x)$ is a polynomial of degree $2^m - 1$ with even coefficients, hence

$$P(x)Q_{2^m}(x) = P(x) + x^{2^m} P(x) + P(x)R(x).$$

In the sum, $P(x)$ and $x^{2^m} P(x)$ have no terms of the same degree, but their coefficients agree, and all coefficients of $P(x)R(x)$ are even, hence (2) holds.

We prove our statement by induction on i_n. If $i_n = 0$ or $i_n = 1$, the statement is obvious. Suppose that (1) holds if $i_n < 2^m$, where m is a positive integer. Starting with this assumption, we show that the statement holds if $2^m \leq i_n < < 2^{m+1}$; this is clearly enough for the proof.

We distinguish two cases. First, assume that the exponents $i_1, i_2, \ldots, i_n$ are between 2^m and 2^{m+1}, that is

$$2^m \leq i_1 < i_2 < \ldots < i_k < 2^{m+1}, \qquad (m \geq 1).$$

In this case

$$Q_{i_1} + Q_{i_2} + \ldots + Q_{i_k} = Q_{2^m} \left(Q_{i_1 - 2^m} + \ldots + Q_{i_n - 2^m} \right).$$

$i_n < 2 \cdot 2^m = 2^{m+1}$, therefore $i_n - 2^m < 2^m$ and so our assumption implies that

$$w \left(Q_{i_1 - 2} + \ldots + Q_{i_n - 2^m} \right) \geq w \left(Q_{i_1 - 2^n} \right),$$

$$w \left(Q_{2^m} \left(Q_{i_1 - 2^m} + \ldots + Q_{i_n - 2^m} \right) \right) = 2w \left(Q_{i_1 - 2^m} + \ldots + Q_{i_n - 2^m} \right) \geq$$

$$\geq 2w \left(Q_{i_1 - 2^m} \right) = w \left(Q_{2^m} Q_{i_1 - 2^m} \right) = w \left(Q_{i_1} \right).$$

Thus we proved the statement.

For the second case, assume that for the exponents $i_1, i_2, \ldots, i_n$

$$i_1 < \ldots < i_{r-1} < 2^m \leq i_r < \ldots < i_n < 2^{m+1}, \qquad (1 < q \leq n)$$

holds. The assumption is, again, that (1) holds for every i_n, where $i_n < 2^m$, thus:

(4) $$w \left(Q_{i_1} + Q_{i_2} + Q_{i_{r-1}} \right) \geq w \left(Q_{i_1} \right).$$

Set

$$Q_{i_1} + Q_{i_2} + \ldots + Q_{i_{r-1}} = a_0 + a_1 + \ldots + a_{2^m - 1} x^{2^m - 1}$$

and use that

$$Q_{i_r} + Q_{i_r + 1} + \ldots + Q_{i_n} = Q_{2^m} \left(Q_{i_r - 2^m} + \ldots + Q_{i_n - 2^m} \right) =$$

$$= \left(1 + x^{2^m} + R(x) \right) \left(Q_{i_r - 2^m} + \ldots + Q_{i_n - 2^m} \right).$$

Let the polynomial in the last parenthesis be:

$$b_0 + b_1 x + \ldots + b_{2^m - 1} x^{2^m - 1},$$

where a_i and b_i are integers. Now,

(5) $$Q_{i_1} + \ldots + Q_{i_{r-1}} + Q_{i_r} + \ldots + Q_{i_n} =$$

(I) $$= a_0 + a_1 x + a_2 x^2 + \ldots + a_{2^m - 1} x^{2^m - 1} +$$

(II) $$+ b_0 + b_1 x + b_2 x^2 + \ldots + b_{2^m - 1} x^{2^m - 1} +$$

(III) $$+ b_0 x^{2^m} + b_1 x^{2^m + 1} + b_2 x^{2^m + 2} + \ldots + b_{2^m - 1} x^{2^{m+1} - 1} + R'(x),$$

where all coefficients of $R'(x)$ are even. Now, it is easy to see that

(6) $$w \left(Q_{i_1} + \ldots + Q_{i_{r-1}} + Q_{i_r} + \ldots + Q_{i_n} \right) \geq w \left(Q_{i_1} + \ldots + Q_{i_{r-1}} \right) =$$

$$= \omega \left(a_0 + a_1 x + \ldots + a_{2^m - 1} x^{2^m - 1} \right),$$

Let us denote the polynomials in the three lines of (5) by I., II. and III. If a_i and b_i are even for some i, then the number of the odd coefficients does not change in (5) compared to I.; if a_i is odd and b_i is even, then since $a_i + b_i$ is odd and the number of odd coefficients remains the same; if a_i is even, b_i is odd, then the number of the odd coefficients is increasing in (5); finally, if a_i and b_i are both odd, then $a_i + b_i$ is even, but in III. b_i is odd, hence the number of odd coefficients is the same. Thus we proved (6), and (6) with (4) implies (1).

Remark. We can visualize the problem using the Pascal triangle. In *Figure 1985/3.1* we presented the mod 2 residues of the shifted Pascal triangle. Thus in the i-th row the 1-s represent the odd and the 0-s the even coefficients of the polynomial $(1+x)^i$. The addition of the polynomial corresponds to the addition of the appropriate rows, coordinatewise. We can reformulate the problem in the

$i =$									
0	1								
1	1	1							
2	1	0	1						
3	1	1	1	1					
2^2	1	0	0	0	1				
5	1	1	0	0	1	1			
6	1	0	1	0	1	0	1		
7	1	1	1	1	1	1	1	1	
2^3	1	0	0	0	0	0	0	0	1

Figure 85/3.1

following way: If we add n rows of the triangle, the number of 1-s in the sum is at least he number of 1-s in the first of the rows.

1985/4. *Given a set M of 1985 distinct positive integers, none of which has a prime divisor greater than 26. Prove that M contains a subset of 4 elements whose product is the 4th power of an integer.*

Solution. The possible prime divisors of the elements of M are

$$2,\ 3,\ 5,\ 7,\ 11,\ 13,\ 17,\ 19,\ 23$$

that is 9 distinct primes. Write the elements of M in the form

$$2^{p_1} \cdot 3^{p_2} \cdot 5^{p_3} \cdot \ldots \cdot 23^{p_9}.$$

Let us assign to every element the 9-tuple $(p_1, p_2, \ldots, p_9)$, where p_i equals 1 if it is odd and 0 if it is even. The number of these 0–1 9-tuples is $2^9 = 512$, hence there are two 9-tuples among any 513 that are identical. Choose two numbers from M, $a_{1,1}$ and $a_{1,2}$, such that the appropriate 9-tuples are the same. Then $a_{1,1}a_{1,2}$ is a perfect square, as all exponents are even. Thus $\sqrt{a_{1,1}a_{1,2}}$ is an integer.

Similarly, from the remaining 1983 numbers we can choose $a_{2,1}$ and $a_{2,2}$ such that $\sqrt{a_{2,1}a_{2,2}}$ is an integer. In a similar fashion we get:

(1) $\qquad \sqrt{a_{1,1}a_{1,2}},\ \sqrt{a_{2,1}a_{2,2}},\ \sqrt{a_{3,1}a_{3,2}},\ \cdots,\ \sqrt{a_{513,1}a_{513,2}}.$

(We choose the last pair from $1985 - 2 \cdot 512 = 961$ numbers.)

The 513 integers listed in (1) have 9 prime divisors. Hence the previous arguments give us two of them, $\sqrt{a_{i,1}a_{i,2}}$, $\sqrt{a_{j,1}a_{j,2}}$ such that the appropriate 0–1 9-tuples are the same. The product of them is a perfect square, hence

$$\sqrt{a_{i,1}a_{i,2}} \cdot \sqrt{a_{j,1}a_{j,2}} = b^2 \qquad (b \text{ integer})$$

and so
$$a_{i,1}a_{i,2}a_{j,1}a_{j,2} = b^4,$$

and this is what we wanted to prove.

Remark. The selection can be made from $2 \cdot 512 + 513 = 1537$ integers.

1985/5. *The circle k_1 with centre O passes through the vertices A and C of the triangle ABC and intersects the segments AB and BC again at distinct points K and N, respectively. The circumcircles k of ABC and k_2 of KBN intersect at exactly two distinct points B and M. Prove that $\angle O_1MB = 90°$.*

Remark. In ABC we have $AB \neq BC$, because otherwise by the symmetry $B = M$ would hold, which is not the case.

First solution. Denote by k the circumcircle of ABC and by O and R its centre and its radius. For k_2, the circumcircle of BKN denote by O_2 and r its centre and radius (see *Figure 1985/5.1*). First we show that BO_2 and AC are perpendicular.

Let P and T denote the second points of intersection of k_2 and the lines BO_2 and AC, respectively. As $ACNK$ is a cyclic quadrangle, $\angle BKN = \gamma$ and $\angle BPN = \gamma$, as they subtend the same arc. Hence the quadrangle $TCNP$ is cyclic, and as by Thales' theorem $\angle BNP$ is a right angle, the opposite angle $\angle PTC = \angle BTC$ is a right angle, too. (We labelled our figure such that γ is acute.)

As $\angle BKN = \gamma$, the included angle of the line e parallel to KN through B and
the line AB is γ, too. Hence
this is the tangent of k and
the radius OB is orthogonal
to e. As e and KN are parallel, OB and KN are perpendicular. O_1O_2 is perpendicular to KN, their chord in
common, hence $OB \parallel O_1O_2$.
OO_1 is orthogonal to their
chord in common, AC (see
Figure 1985/5.2).

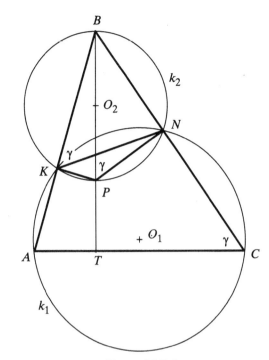

Figure 85/5.1

But, BO_2 and AC are
perpendicular, hence the opposite sides of the quad-
rangle OBO_2O_1 are parallel, so $OB = O_1O_2 = R$ and
$OO_1 = BO_2 = r$. As O_2P is
parallel and equal to OO_1,
OO_1PO_2 is a parallelogram.

Denote by M' the sec-
ond point os intersection of
O_1P and k_2. In the quadrangle O_1OO_2M' the sides
O_1M' and OO_2 are parallel,
$O_1M' > OO_2$, $OO_1 = O_2M' =$
$= r$, hence O_1OO_2M' is a
symmetric trapezium and so
the lengths of its diagonals are
the same: $O_1O_2 = OM' = R$.
This implies that M' is on
k, the circumcircle of ABC
thus $M \equiv M'$. In k_2 the interval PB is a diameter, hence
$\angle PMB$ is a right angle, thus
$\angle O_1MB = \angle PMB = 90°$, and
we proved our statement.

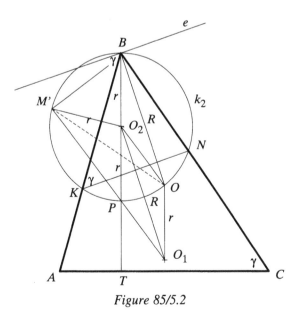

Figure 85/5.2

Second solution. Let r_1 denote the radius of k_1. The radical axes of the pairs of the circle, the lines AC, KN, BM of the common chords of the circles k, k_1, k_2 concur in Q the so called radical centre ([6]). This point exists as $AB \neq BC$; we may assume that A separates Q and C (*Figure 1985/5.3*).

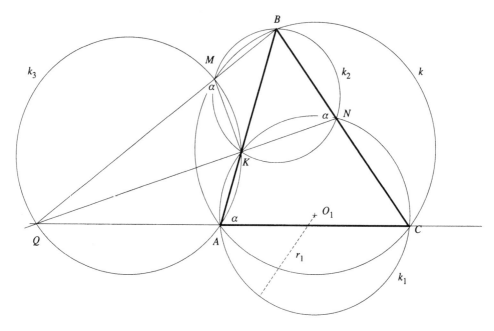

Figure 85/5.3

Since $ACNK$ and $BMKN$ are cyclic quadrangles, $\angle KNB = \alpha$ and $\angle QMK = \alpha$, and so $AKMQ$ is cyclic, too, as the sum of the angles at A and M is $180°$. Let k_3 be the circumcircle of $AKMQ$.

By the power of the point theorem:

(1) $$QO_1^2 - r_1^2 = QA \cdot QC = QM \cdot QB = QM(QM + MB).$$

and

(2) $$BO_1^2 - r_1^2 = BC \cdot BN = BM \cdot BQ = MB(QM + MB).$$

subtracting (1) from (2) we get

$$QO_1^2 - BO_1^2 = (QM - MB)(QM + MB) = QM^2 - MB^2.$$

which implies that BQ is perpendicular to MO_1-re, and this is what we wanted to prove.

Remarks. 1. At the last step of the second solution we used that if M is a point of the line AB and P is an arbitrary point, then PM is orthogonal to AB if and only if

(3) $$AM^2 - BM^2 = AP^2 - BP^2.$$

Indeed,

$$AP^2 - BP^2 = \overrightarrow{AP}^2 - \overrightarrow{BP}^2 = \left(\overrightarrow{MA} - \overrightarrow{MP}\right)^2 - \left(\overrightarrow{MB} - \overrightarrow{MP}\right)^2 =$$

$$= \overrightarrow{MA}^2 - \overrightarrow{MB}^2 + 2\overrightarrow{MP}\left(\overrightarrow{MB} - \overrightarrow{MA}\right) = AM^2 - BM^2 + 2\overrightarrow{MP}\cdot\overrightarrow{AB}.$$

Hence (3) holds if and only if $\overrightarrow{MP}\cdot\overrightarrow{AB} = 0$, that is if PM and AB are perpendicular.

2. This problem has an interesting connection to the theory of conic sections. If a parabola touches the sides of the cyclic quadrangle $ACNK$ then its focus M lies on the line connecting B and Q, the points of intersection of the opposite sides of the quadrangle. Moreover, BQ is perpendicular to the line through M and the centre of the quadrangle (for further details see [36]).

1985/6. *For every real number x_1 construct the sequence $x_1, x_2, \ldots, x_n$, where*

(1) $$x_{n+1} = x_n \left(x_n + \frac{1}{n}\right)$$

for every $n \geq 1$. Prove that there exists exactly one value of x_1 which gives

$$0 < x_n < x_{n+1} < 1$$

for every positive n.

Solution. Fist we show that there is at most one x_1 that satisfies the conditions. By the assumptions $\{x_i\}$ is strictly increasing and bounded, hence it has a limit, x. By (1) it satisfies the equation

$$x = x\left(x + \lim_{n\to\infty}\frac{1}{n}\right)$$

$$x = x^2$$

thus $x = 1$.

Now, assume that there are two appropriate starting values, x_1 and y_1, where $x_1 < y_1$. We prove by induction that $x_n < y_n$, for every positive integer n. By the assumptions $x_{n-1} < y_{n-1}$.

$$y_n - x_n = y_{n-1}^2 + \frac{y_{n-1}}{n-1} - x_{n-1}^2 - \frac{x_{n-1}}{n-1} =$$

(2) $$= \left(y_{n-1} - x_{n-1}\right)\left(x_{n-1} + y_{n-1} + \frac{1}{n-1}\right) > 0.$$

By the convergence of the sequences there is an index N such that for $N < n$ the elements x_n and y_n are greater than $\frac{3}{4}$, hence from (2) we get:

$$y_n - x_n > \frac{3}{2}(y_{n-1} - x_{n-1})$$

$$y_{n+1} - x_{n+1} > \frac{3}{2}(y_n - x_n) > \left(\frac{3}{2}\right)^2 (y_{n-1} - x_{n-1})$$

$$\vdots$$

$$y_{n+k} - x_{n+k} > \left(\frac{3}{2}\right)^{k+1} (y_{n-1} - x_{n-1}).$$

Since $y_{n-1} - x_{n-1}$ is fix, the r.h.s. goes to infinity, giving a contradiction. Thus there is a single value for x_1.

We show that such an x_1 exists.

Construct the elements $x_1, x_2, \ldots$ as a function of $x = x_1$.

$$x_2 = f_2(x) = x^2 + x, \qquad f_3(x) = f_2(x)^2 + \frac{f_2(x)}{2} = x^4 + 2x^3 + \frac{3x^2}{2} + \frac{x}{2},$$

(3)
$$f_n(x) = f_{n-1}(x) \left(f_{n-1}(x) + \frac{1}{n-1} \right).$$

As $f_2(x)$ is a strictly increasing function of x, the same holds for $f_i(x)$ the other functions of the sequence. (3) implies $f_i(0) = 0$ and $\lim_{x \to \infty} f_i(x) = \infty$.

Define the sequences

$$a_1, a_2, \ldots, a_n, \ldots \qquad \text{and} \qquad b_1, b_2, \ldots, b_n, \ldots$$

as follows:

$$f_n(a_n) = 1 - \frac{1}{n}, \qquad f_n(b_n) = 1.$$

As $f_i(x)$ strictly increasing and continuous, it attains every value on $[0, \infty[$ and so a_n and b_n are uniquely determined.

The sequence a_i is strictly increasing. Indeed,

$$f_{n+1}(a_{n+1}) - f_{n+1}(a_n) = 1 - \frac{1}{n+1} - f_n(a_n) \left(f_n(a_n) + \frac{1}{n} \right) =$$

$$= 1 - \frac{1}{n+1} - \left(1 - \frac{1}{n}\right) \left(1 - \frac{1}{n} + \frac{1}{n}\right) = \frac{1}{n(n+1)} > 0,$$

hence by the monotonicity of $f_{n+1}(x)$ we have $a_{n+1} > a_n$.

The sequence b_i is strictly decreasing:

$$f_{n+1}(b_{n+1}) - f_{n+1}(b_n) = 1 - f_n(b_n) \left(f_n(b_n) + \frac{1}{n} \right) = -\frac{1}{n} < 0,$$

thus $b_{n+1} < b_n$.

The $[a_i, b_i]$ intervals are nested, because by definition $f_n(a_n) < f_n(b_n)$ and so by the monotonicity of $f_n(x)$ we have $a_n < b_n$ for every n. If there were a_k, b_i, such that $b_i < a_k$ and $i < k$, then it would imply $a_k > b_i > b_{i+1} > \ldots > b_k$, contradicting the inequality $a_k < b_k$. We get a similar contradiction starting with the $i > k$ assumption.

Thus the $[a_i, b_i]$ intervals have a common point c, such that for every i the inequality $a_i < c < b_i$ holds, as $a_i < a_{i+1} \leq c \leq b_{i+1} < b_i$.

We show that choosing $x_1 = c$ the sequence satisfy the conditions.

$f_n(a_n) < f_n(c)$, because $a_n < c$ and so

(4) $\qquad 1 - \dfrac{1}{n} < f_n(c) \qquad$ and so $\qquad f_n(c) - \dfrac{f_n(c)}{n} < f_n^2(c),$

(5) $\qquad f_n(c) < f_n^2(c) + \dfrac{f_n(c)}{n} = f_{n+1}(c).$

Moreover, $c < b_{n+1}$ implies

(6) $\qquad f_{n+1}(c) < f_{n+1}(b_{n+1}) = 1,$

furthermore (4), (5) and (6) mean that

$$0 < f_n(c) < f_{n+1}(c) < 1,$$

hence for the sequence starting with c

$$0 < x_n < x_{n+1} < 1$$

holds, and we proved our statement.

Remark. c can be calculated: $c = 0,4465349\ldots$

1986.

1986/1. *Let d be any positive integer not equal to 2, 5 or 13. Show that one can find distinct a, b in the set $\{2, 5, 13, d\}$ such that $ab - 1$ is not a perfect square.*

Solution. If a and b are among the first three numbers of the set, then the values of $(ab - 1)$, 9, 25, 64 are perfect squares. We only have to check whether we can make perfect squares with d. These numbers are $2d - 1$, $5d - 1$, $13d - 1$. We have to show that they are not all perfect squares.

Assume to the contrary that there are positive integers a, b, c such that

(1) $\qquad\qquad\qquad 2d - 1 = a^2,$

(2) $\qquad\qquad\qquad 5d - 1 = b^2,$

(3) $\qquad\qquad\qquad 13d - 1 = c^2.$

(1) implies that a is odd, hence its square is of the form $a^2 = 8k + 1$,

$$2d - 1 = 8k + 1,$$

(4) $\qquad\qquad\qquad d = 4k + 1,$

hence d is odd. From (2) and (3) we see that b and c are even. Let $b = 2b_1$, $c = 2c_1$; Now, the difference of the appropriate sides of (3) and (2) reads

$$8d = c^2 - b^2 = 4(c_1^2 - b_1^2) = 4(c_1 + b_1)(c_1 - b_1),$$

(5) $\qquad\qquad\qquad 2d = (c_1 + b_1)(c_1 - b_1).$

As the sum and the difference of two numbers have the same parity, both $c_1 + b_1$ and $c_1 - b_1$ are even, since their product is even. Thus the l.h.s. of (5) is the product of two even numbers, hence divisible by 4; let us denote it by $4r$:

$$2d = 4r,$$

$$d = 2r$$

But this implies that d is even contradicting that it was odd. Hence the equations (1)–(3) have no solution.

1986/2. *Given a point P in the plane of the $A_1A_2A_3$ triangle. Define $A_s = = A_{s-3}$ for $s \geq 4$.*

Construct a series of points P_0, P_1, P_2, ... such that P_{k+1} is the image of P_k under a rotation with centre A_{k+1} through an angle $-120°$ ($k = 0, 1, 2, ...$).

Prove that if $P_{1986} = P_0$, then the triangle $A_1A_2A_3$ is equilateral.

First solution. The problem can be rewritten as: compose the rotations with centres A_1, A_2, A_3 through an angle $-120°$ and repeat it 662 times ($662 \cdot 3 = 1986$). We have to prove that if P_0 is a fix point of this transformation, then the triangle $A_1A_2A_3$ is equilateral.

The composition of three rotations with an angle of $-120°$ is a translation since it maps every vector to itself. Let $\mathbf{v}$ denote the vector of the translation. The composition of 662 translations is a translation by $662\mathbf{v}$ that has the fixpoint P_0. Thus this transformation is the identity:

$$662\mathbf{v} = \mathbf{0}, \qquad \mathbf{v} = \mathbf{0}.$$

Apply these three rotations to A_1. The first rotation fixes A_1, the second maps it to a point A_1' and the third takes it back to A_1. (see *Figure 1986/2.1*). Thus $A_1A_2A_1'A_3$ is a quadrangle, where $\angle A_1A_2A_1' = \angle A_1'A_3A_1 = 120°$, $A_1A_2 = = A_2A_1'$, $A_1'A_3 = A_3A_1$, thus it is a deltoid, hence the diagonal A_2A_3 bisects the angle. But then $\angle A_1A_2A_3 = \angle A_2A_3A_1 = 60°$, therefore the triangle is equilateral.

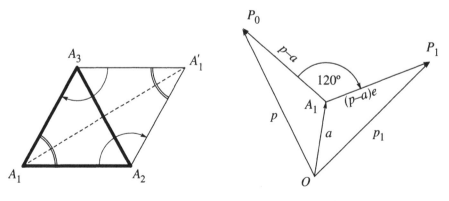

$$\text{Figure 86/2.1} \qquad\qquad\qquad\qquad \text{Figure 86/2.2}$$

Second solution. Let us consider our configuration on the complex plane with the centre of gravity as 0. Let a, b, c denote the numbers of the vertices.

By the choice of the origin

(1) $$a+b+c=0.$$

Let $e=\cos(-120°)+i\sin(-120°)=\cos 120° - i\sin 120°$; now, $e^3=1$ and $e^3-1=(e-1)(e^2+e+1)=0$, which implies $e^2+e+1=0$ and so

(2) $$e^2=-e-1.$$

The geometric meaning of the multiplication by e is a rotation by $-120°$ about the origin. Let p denote the complex number belonging to P_0, and p_i the one to P_i $(i=1, 2, \ldots)$

$$p_1 = a+(p-a)e = a+pe-ae.$$

Similarly, we can calculate the result after the rotation about the points A_2 and A_3:

$$p_2 = b+(p_1-b)e = b+(a-b+pe-ae)e = pe^2-ae^2-be+ae+b.$$
$$p_3 = c+(p_2-c)e = p+(a-b)e^2+(b-c)e+c-a.$$

Using (2) and (1) we get

(3) $$p_3 = p+3(be-a).$$

So the composition of the rotations about the points A_1, A_2, A_3 is expressed with (3). Define $be-a=v$.

$$p_3 = p+3v, \quad p_6 = p+6v, \quad p_9 = p+9v, \quad \ldots, \quad p_{1986} = p+1986v = p.$$

Hence $1986v=0$, $v=0$, $be=a$, (1) and (2) imply

$$be=a=-b-c, \qquad c=b(-e-1)=be^2.$$

This means that rotating A_2 through $120°$ about 0 we get A_1, and through $240°$ we get A_3, thus the triangle $A_1A_2A_3$ is equilateral.

Remarks. 1. In the solution the only thing we used about 1986 is that it is divisible by 3.

2. Our solution was based on Napoleon's theorem: The centres of the equilateral triangles constructed over the sides of an arbitrary triangle (outside) form an equilateral triangle. The centres of the equilateral triangles constructed over the sides of the $P_0P_1P_2$ triangle are A_1, A_2, A_3. (see remark after 1975/3).

1986/3. *To each vertex of a regular pentagon an integer is assigned, so that the sum of all five numbers is positive. If three consecutive vertices are assigned the numbers x, y, z respectively, and $y < 0$, then the following operation is allowed:*

x, y, z are replaced by $x+y$, $-y$, $z+y$ respectively. Such an operation is performed repeatedly as long as at least one of the five numbers is negative. Determine whether this procedure necessarily comes to an end after a finite number of steps.

Solution. We shall base our proof on the observation that during an allowed operation the sum of the numbers remains the same.

Denote the five numbers by: x_1, x_2, x_3, x_4 and x_5, their sum by S. Consider the following function of the variables:

$$f(x_1, x_2, x_3, x_4, x_5) =$$
$$= (x_1 - x_4)^2 + (x_2 - x_5)^2 + (x_3 - x_1)^2 + (x_4 - x_2)^2 + (x_5 - x_3)^2.$$

We show that after performing an allowed operation for the $\{x_i\}$-s, the value of the function is decreasing. As the range of f is the set of the non-negative integers, and at every operation its value is decreased by at least 1, the procedure terminates in finitely many steps. Hence it is enough to prove that f is strictly increasing.

Let $X(x_1, x_2, x_3, x_4, x_5)$ be an appropriate 5–tuple such that one of the numbers is negative. As f is symmetric, we may assume that $x_3 < 0$.

After an allowed operation X transforms to $Y(x_1, x_2 + x_3, -x_3, x_3 + x_4, x_5)$. We have to show that $f(X) > f(Y)$, that is, $f(Y) - f(X) < 0$.

$$f(Y) - f(X) = (x_1 - x_3 - x_4)^2 + (x_2 + x_3 - x_5)^2 + (-x_3 - x_1)^2 + (x_4 - x_2)^2 +$$
$$+ (x_5 + x_3)^2 - (x_1 - x_4)^2 - (x_2 - x_5)^2 - (x_3 - x_1)^2 - (x_4 - x_2)^2 - (x_5 - x_3)^2 =$$
$$= 2x_1 x_3 + 2x_2 x_3 + 2x_3^2 + 2x_3 x_4 + 2x_3 x_5 = 2x_3 S < 0,$$

hence we proved our statement.

Remarks. 1. You can replace f by any function that decreases after an allowed operation.

Géza Kós produced a fancy function:

$$g(x) = 2(x_1 x_2 + x_2 x_3 + x_3 x_4 + x_4 x_5 + x_5 x_1) +$$
$$+ 3(x_1 x_3 + x_2 x_4 + x_3 x_5 + x_4 x_1 + x_5 x_2).$$

g is strictly increasing, but its value is bounded from above: $g(X) \leq S^2$. This is a consequence of

$$g = S^2 - \frac{f}{2}.$$

The following $c(X)$ works well, too:

$$c(X) = x_1^2 + x_2^2 + x_3^2 + x_4^2 + x_5^2 + (x_1 + x_2 + x_3)^2 + (x_2 + x_3 + x_4)^2 +$$
$$+ (x_3 + x_4 + x_5)^2 + (x_4 + x_5 + x_1)^2 + (x_5 + x_1 + x_2)^2.$$

$c = f + 2S^2$ shows its monotonity.

2. The following function produced by an American contestant won a special prize:

$$a(x_1, x_2, x_3, x_4, x_5) = |x_1| + |x_2| + |x_3| + |x_4| + |x_5| +$$
$$+ |x_1 + x_2| + |x_2 + x_3| + |x_3 + x_4| + |x_4 + x_5| + |x_5 + x_1| +$$
$$+ |x_1 + x_2 + x_3| + |x_2 + x_3 + x_4| + |x_3 + x_4 + x_5| + |x_4 + x_5 + x_1| + |x_5 + x_1 + x_2| +$$
$$+ |x_1 + x_2 + x_3 + x_4| + |x_2 + x_3 + x_4 + x_5| + |x_3 + x_4 + x_5 + x_1| +$$
$$+ |x_4 + x_5 + x_1 + x_2| + |x_5 + x_1 + x_2 + x_3|.$$

We may assume that $x_2 < 0$. After the allowed operation:

$$a(X) - a(Y) = |x_3 + x_4 + x_5 + x_1| - |x_2 + 2x_2 + x_3 + x_4 + x_5|$$
$$= |S - x_2| - |S + x_2| > 0,$$

thus $a(X)$ is strictly increasing. The advantage of this function is that it is easy to handle.

3. The statement remains true for an n-gon instead of a pentagon.

1986/4.   *Let A, B be adjacent vertices of a regular n-gon $(n \geq 5)$ with centre O. A triangle XYZ, which is congruent to and initially coincides with OAB, moves in the plane in such a way that Y and Z each trace out the whole boundary of the polygon, with X remaining inside the polygon. Find the locus of X.*

Solution.   Introduce the following notations: Let 1 be the radius of the incircle, $\angle AOB = 2\pi/n = 2\alpha$, so $\alpha = \pi/n$; $\angle OAB = \angle OBA = \beta = \dfrac{\pi}{2} - \alpha$.

The rotation with centre O and by angle 2α maps the polygon to itself, so it is enough to examine the movement by the AB and BC sides. Rotating the points X obtained by angles 2α, 4α, $\ldots$, $(n-1)2\alpha$ we get the locus (see *Figure 1986/4.1*).

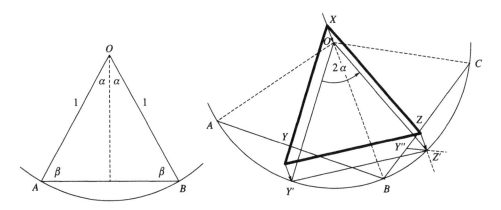

Figure 86/4.1

Let A, B, C be three consecutive vertices of the polygon. (in counterclockwise order). If Y is on the side AB, then Z lies on BC, because the inner points of the other edges are farther from AB.

If $Y = A$, then $Z = B$ and so X coincides with O; The same holds for $Y = B$. Assume that Y is an inner point of AB. Translate the triangle XYZ by $\overrightarrow{XO}$, the vertex X is mapped to O, moreover Y to Y' and Z to Z'. As the distance of Y' and Z' from O equal 1, they lie on the circumcircle of the polygon and $\overrightarrow{XO} = \overrightarrow{YY'} = \overrightarrow{ZZ'}$. Now, rotate the points A, B, Y, Y' about O (counterclockwise) through angle 2α their images are B, C, Y'', Z', where Y'' is an inner point of BC.

Since $YY' = ZZ' = Y''Z'$, the triangle $ZY''Z'$ is isosceles and its interior angle is 2α, because $Y''Z'$ was rotated by 2α. Now, the isosceles triangles $ZY''Z'$ and BCO are similar and they lie on the different sides of he line BC. Hence their sides are parallel, and so OB is parallel to ZZ', but then $\overrightarrow{ZZ'} = \overrightarrow{XO}$ is parallel to OB, too. This means that X is on the line BO, on the ray from O not containing B.

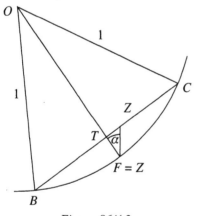

Figure 86/4.2

Our result shows that while Y moves from A to B, X passes along a segment, starting at O, to its farthest location, Q, and returns to O. We determine the exact location of Q. Observe that the distance from X to O equals ZZ' and this distance is the largest if the $ZY''Z'$ triangle is the largest. This latter is the largest if the altitude of the triangle is the largest. This altitude is the distance of Z' from the line BC, which is the largest if Z' is the midpoint F of the arc BC. Now it is easy to calculate the length of $OQ = ZF$ (see *Figure 1986/4.2*).

Let T denote the foot of the altitude from O in the OBC triangle. From the similarity of the right triangles FZT and OCT we get

$$\frac{ZF}{FT} = \frac{1}{OT}.$$

Hence

$$ZF = \frac{FT}{OT} = \frac{1 - OT}{OT} = \frac{1 - \cos\alpha}{\cos\alpha} = OQ.$$

It is remained to show that every point of the OQ segment is the vertex of a moving triangle. Indeed, translate BC by $\overrightarrow{XO}$, the image of the segment intersects the BC arc in two points (or touches at the midpoint). Let Z' be one of the points of intersection. The line through Z' parallel to OB intersects BC

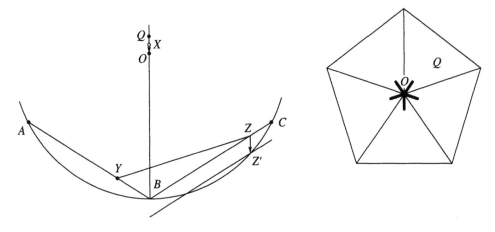

Figure 86/4.3

in the point Z. There is a single point Y on the side AB such that its distance from Z is $AB = BC$. The distance of vertex X of the triangle with base YZ congruent to OAB from O — as we saw before — equals to ZZ' hence the point is X (see *Figure 1986/4.3*).

Thus the locus is the union of n segments starting at O such that their other vertices form a regular n-gon.

Remarks. 1.In our solution we used visual arguments. These can be formulated more precisely.

2. The problem belongs to the subject of kinetic geometry that has several technical applications. We present a famous theorem of kinetic geometry:

Let be given two covering planes. Fix one of them and move the other one in the plane. Every point of the moving plane describes a curve. At every stage of the movement the normal vectors constructed to the points of the plane intersect in a point or they are parallel. This point of intersection is called the momentary centre of the movement (*Figure 1986/4.4*).

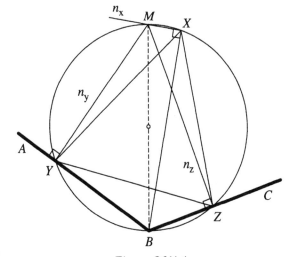

Figure 86/4.4

In our problem the plane of the polygon is the fix plane and the moving plane contains the XYZ triangle. During the movement the points Y and Z describe a segment, the point of intersection of the normal vectors $\mathbf{n}_Y$, $\mathbf{n}_Z$ intersect in M, the momentary centre. As the quadrangle $MYBZ$ is a cyclic quadrangle, both M and X are on the circumcircle of XYZ, where MB is a diameter, hence Y and Z subtend a right angle. M lies on the normal vector of the orbit of X, $\mathbf{n}_X$, too. Thales' theorem implies that BX is perpendicular to MX, thus BX is the tangent of the orbit of X; hence X describes a curve such that all tangents go through a fixed point, B. It can be shown that only a line bears this property, thus X moves on a line.

1986/5. *Find all functions f defined on the non-negative reals and taking non-negative real values such that:*

(a) $f(x \cdot f(y)) \cdot f(y) = f(x+y)$ *for every non negative x and y;*

(b) $f(2) = 0$;

(c) $f(x) \neq 0$, *if $0 \leq x < 2$.*

Solution. First assume that $x \geq 2$ and let $y = 2$; by (a):

$$f((x-2)f(2)) \cdot f(2) = f(x-2+2) = f(x),$$

(b) implies $f(2) = 0$ and so for every $x \geq 2$

$$(1) \qquad\qquad f(x) = 0,$$

and this holds only in case $x \geq 2$ according to (c).

Now, let $0 \leq y < 2$. The l.h.s of (a) is 0, if $xf(y) \geq 2$, that is

$$(2) \qquad\qquad x \geq \frac{2}{f(y)};$$

the r.h.s. equals 0 if $x + y \geq 2$, thus

$$(3) \qquad\qquad x \geq 2 - y.$$

This implies, that under the given conditions the equality

$$(4) \qquad\qquad \frac{2}{f(y)} = 2 - y$$

holds. Indeed, if it does not hold for some y, than there is a smaller number on the l.h.s. of (4) than on the r.h.s. and hence there were an x, such that

$$(5) \qquad\qquad \frac{2}{f(y)} < x < 2 - y$$

held. But now, (2) implies that there is 0 on the l.h.s. of (a), moreover (c) implies that the r.h.s. is not 0, that gives a contradiction. Similarly, it can be argued against the opposite inequality in (5). Thus for every $0 \leq y < 2$ the equation (4) holds:

(6) $$f(y) = \frac{2}{2-y}, \qquad (0 \le y < 2).$$

Thus the only function satisfying the conditions (a), (b), (c) is the following:

$$f(x) = \begin{cases} \dfrac{2}{2-x}, & \text{if } 0 \le x < 2, \\ 0, & \text{if } x \ge 2. \end{cases}$$

It remains to show that this function satisfies the conditions. (b) and (c) are obviously true, if $y \ge 2$, both sides of (a) is 0. Thus it is enough to consider the case $0 \le y < 2$.

First, assume that $x + y \ge 2$. Hence the r.h.s. is 0. Then $x \ge 2 - y$, and

$$x \cdot f(y) = \frac{2x}{2-y} \ge \frac{2(2-y)}{2-y} = 2,$$

thus $f(xf(y)) = 0$, and so the l.h.s. is 0, too.

If $x + y < 2$, then $x < 2 - y$, and

$$x \cdot f(y) = \frac{2x}{2-y} < \frac{2(2-y)}{2-y} = 2,$$

and so

$$f(x \cdot f(y)) \cdot f(y) = \frac{2}{2 - \frac{2x}{2-y}} \cdot \frac{2}{2-y} = \frac{2(2-y) \cdot 2}{(4 - 2y - 2x)(2-y)} =$$

$$= \frac{2}{2 - (x+y)} = f(x+y),$$

so $f(x)$ satisfies (a).

1986/6. *Given a finite set of points in the plane, each with integer coordinates. Is it always possible to colour the points red or white so that for any straight line L parallel to one of the coordinate axes the difference (in absolute value) between the numbers of white and red points on L is not greater than 1?*

First solution. Let us call the lines parallel to the axes containing lattice points lattice lines and denote by M the set of the given lattice points. There are two lattice lines through every point of M. We prove that there is a colouring of M such that for any lattice line the difference between the numbers of white and red points is not greater than 1.

For simplicity we may assume that both coordinates of the points in M are positive.

If there is a point P in M, such that neither of the two lattice lines through P contains a point of M different from P, then we colour P white. From now, we omit these points.

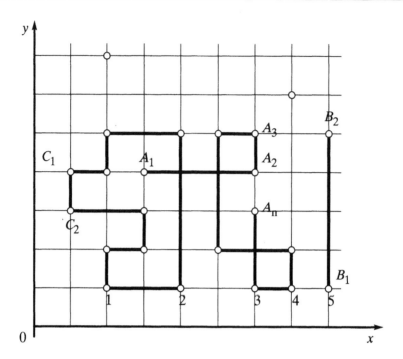

Figure 86/6.1

If an arbitrary lattice line contains more than 1 point of M, number the points from left to right (bottom to top) by the integers 1, 2, ... Connect on each line the pairs of points:

$$(1,2), \quad (3,4), \quad (5,6), \quad \ldots, \quad (2k-1, 2k), \quad \ldots$$

The only points that are not connected to any other points are the "last" points from the axes in case the appropriate lines contain odd number of points. Now the points of M form a graph such that the degrees of the points are 0, 1 or 2 (see *Figure 1986/6.1*).

This graph is the disjoint union of open or closed polygonal paths that are the concatenations of the line segments defined above.

Choose a point A_1 of degree 1; it is connected to a unique lattice point A_2. If A_2 is of degree 1, the path terminated; if not, it is uniquely continued to A_3. We can proceed until we arrive to A_n, a point of degree 1. If these points $A_1, A_2, \ldots, A_n$ do not cover M, choose B_1, a point of degree 1 and construct a polygonal path starting at B_1. We continue the procedure until there are no points of degree 1 left.

Now, omit these polygonal paths from M, and if there is a point of degree 2, e.g. C_1, then we can construct a closed polygonal path returning to C_1. Thus we connected the points of M by open and closed polygonal paths. Two paths cannot share a point in common, because that was of degree at least 3, contradiction.

We colour the points path by path in the following way: choose a path and an orientation of the path. Colour the points one after another on the path red and white, alternating. If the polygonal path is closed, it has even many vertices (see our remark), hence the endpoints of every segment are of different colours.

We prove that this colouring satisfies the conditions of the problem. The endpoints of every segment are coloured with different colours, and on every lattice line there is at most one point that is not the endpoint of a segment on the line, hence the difference of the numbers of the red and white points is at most 1.

Second solution. We prove the statement by induction. For small numbers (1, 2, 3, 4 points) the statement is obvious. Assume that for the numbers smaller than n the colouring exists. We prove that there is an appropriate colouring for n points.

If there is a point P such that no lattice line through P contains another point, then the remaining $n-1$ points can be coloured by the conditions and colour P arbitrarily.

If there is no point with the above property, choose two points of M, A and B from the same line. We distinguish two cases:

1. There is no other point of M on the perpendiculars through A and B;

2. At least one of the perpendiculars contain another lattice point.

In case 1. the $n-2$ points distinct to A and B can be coloured by the assumption; moreover let A be white and B red. Now, the number of red and white points remained the same in the lines distinct from AB, and on the AB line the difference is the same.

In case 2. there is a point C of M on the perpendicular through B. Let X be the point completing the triple A, B, C to a rectangle: X is in M or not (see *Figure 1986/6.2*).

If X is among the given points, omit A, B, C and X. The remaining $n-4$ points can be coloured by the assumptions and colour A and C red, B and X white. Thus the difference of the number of red and white points remained the same.

If X is not in M, omit A, B, C and include X to the set. The set of size $n-2$ can be coloured. Now, colour A and C the same

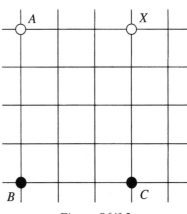

Figure 86/6.2

colour as X and B the opposite colour, then omit X. The colouring given this way does not change the difference of the number of red and white points. Thus we proved that the colouring exists.

Remark. In the first solution we used that if every consecutive pairs of sides of a closed polygonal path is perpendicular then it has even many vertices.

Let us direct our vectors in some orientation. The angle between consecutive vector pairs is $90°$ or $3 \cdot 90°$. The sum of these angles is a multiple of $360°$. This is an even multiple of $90°$; only even many odd numbers can add up to be even, so the number of the vectors is even.

1987.

1987/1. *Let $p_n(k)$ be the number of permutations of the set $S = \{1, 2, \ldots \ldots, n\}$ $(n \geq 1)$ which have exactly k fixed points. Prove that the sum from $k = 0$ to n of $kp_n(k)$ is $n!$.*

First solution. Enumerate the altogether $n!$ permutations of the n numbers and mark the fixed points on the list i.e., an occurrence of 5, for example, is marked whenever it is in the fifth column:

$$\textcircled{1} \; \textcircled{2} \; \textcircled{3} \; \textcircled{4} \; \textcircled{5} \; \ldots \; \textcircled{n}$$

$$2 \quad 1 \quad \textcircled{3} \quad 7 \quad \textcircled{5} \; \ldots \; 9$$

$$\textcircled{1} \; \textcircled{2} \; 5 \; \textcircled{4} \; 7 \; \ldots \; \textcircled{n}$$

There are k numbers marked this way in each row where there are k fixed points. For the $p_n(k)$ permutations of the problem there are, altogether, $p_n(k)$ numbers marked in the rows containing k fixed points for every k. Hence the sum

$$\sum_{k=0}^{n} kp_n(k)$$

is the total number of marked elements in the $n!$ rows.

Let us also count the marked elements column wise. Every number, from 1 to n, occurs, by symmetry, the same number of times in each column, namely $\dfrac{n!}{n} = (n-1)!$ times. Hence i occurs this many times in the ith column, in particular. Therefore, there are $(n-1)!$ marked elements in each column. Being so there are $n(n-1)! = n!$ marked numbers in the array altogether, which, when compared to the previous result yields the claim:

$$\sum_{k=0}^{n} kp_n(k) = n!$$

Second solution. Let's try to find a relation between $p_n(k)$, the number of k-fixed n-permutations, for short, and that of the $(k-1)$-fixed $(n-1)$-permutations, $p_{n-1}(k-1)$.

126

Keeping any one of the n numbers fixed at its position, the $(k-1)$-fixed permutations of the remaining $n-1$ elements — there are $p_{n-1}(k-1)$ of them — form, with the fixed element, a k-fixed n-permutation. There are n ways to choose the fixed element so we obtain $n \cdot p_{n-1}(k-1)$ k-fixed n-permutations and every one of them has been checked this way. Each k-fixed n permutation, on the other hand, has been listed exactly k times: indeed, consider the 3-fixed permutation 1 4 3 2 5 of the elements 1 2 3 4 5, for example. According to the tally above this very permutation can be obtained by fixing 1 at the first position and preparing the 2-fixed permutations of the remaining 4 elements. Since we can also start by fixing 3 or 5 there are $3 \cdot P_4(2)$ of them, indeed. In general

$$(1) \qquad kp_n(k) = np_{n-1}(k-1).$$

Having respectively plugged 1, 2, ...,n in (1) for k and summing up the arising equalities yields

$$(2) \qquad \sum_{k=1}^{n} kp_n(k) = \sum_{k=1}^{n} np_{n-1}(k-1) = n\sum_{k=1}^{n} p_{n-1}(k-1).$$

Observe now that $0 \cdot p_n(0)=0$ and, as the sum of the numbers of 0-fixed, 1-fixed,...,$(n-1)$-fixed $(n-1)$-permutations, respectively, $\sum_{k=1}^{n} p_{n-1}(k-1)$ is equal to $(n-1)!$ Hence (2) can be written as

$$\sum_{k=0}^{n} kp_n(k) = n(n-1)! = n!$$

and the proof is complete.

Third solution. Denote, for brevity, the number of those n-permutations where there are no fixed points at all by $p(n)$, that is let $p(n) = p_n(0)$.

Fixing k elements out of n, each at its own position and permuting, with no fixed points this time, the remaining $(n-k)$ ones yields k-fixed permutations and clearly $p(n-k)$ of them. Since there are $\binom{n}{k}$ ways to fix k numbers the total number of k-fixed n-permutations is

$$(1) \qquad p_n(k) = \binom{n}{k} p(n-k).$$

Given that $\binom{n}{k} = \frac{n}{k}\binom{n-1}{k-1}$,

$$(2) \qquad k \cdot p_n(k) = n\binom{n-1}{k-1} p(n-k),$$

Since $0 \cdot p_n(0) = 0$ this implies

$$\sum_{k=0}^{n} k p_n(k) = \sum_{k=1}^{n} n \binom{n-1}{k-1} p(n-k) = n \sum_{k=1}^{n} \binom{n-1}{k-1} p(n-k).$$

According to (1) the sum left to be computed is

$$p_{n-1}(n-1) + p_{n-1}(n-2) + \ldots + p_{n-1}(0)$$

and this is the sum of the number of $(n-1)$-fixed, $(n-2)$-fixed, $\ldots$, 0-fixed $(n-1)$-permutations, respectively; the rest follows as before.

Remarks. 1. The origin of the problem is the infamous 'problem of the confounded letters' from the first part of the 18th century. It goes like this: each one of n letters is put into one of n addressed envelopes; what is the probability that each letter goes to wrong destination. Labelling the letters by numbers the number of favourable outcomes is exactly the number of 0-fixed permutations of n elements, the number of so called *derangements*. N. *Bernoulli* (1687–1759) has found the following formula:

$$(3) \qquad p(n) = p_n(0) = n! \left(\frac{1}{2!} - \frac{1}{3!} + \frac{1}{4!} - \ldots + \frac{(-1)^n}{n!} \right).$$

(3), by the way, satisfies the recurrence $p(n) = (n-1)(p(n-1) + p(n-2))$ and it can also be used to complete the argument of the *Third solution.*

2. The original proposal also required the proof of

$$\sum_{k=0}^{n} (k-1)^2 p_n(k) = n!;$$

this can be proved by any one of the outlined methods.

1987/2. *In an acute-angled triangle ABC the interior bisector of angle A meets BC at L and meets the circumcircle of ABC again at N. From L perpendiculars are drawn to AB and AC, with feet K and M respectively. Prove that the quadrilateral $AKNM$ and the triangle ABC have equal areas.*

First solution. The idea of the proof is to show that the area left when removing the triangle BNC from the quadrilateral $ABNC$ is equal to what is left of the very same quadrilateral if the two triangles, KBN and MNC are removed. For this it is enough to show that $[BNC] = [KBN] + [MNC]$ and this is going to be done by dividing the triangle BNC in such a way that the respective parts are of the same area as the triangles KBN and MNC (*Figure 87/2.1*).

First of all, K and M are interior to the respective sides AB and AC because the triangle ABC is acute. Denote the second intersection of the line BC and the circumcircle of the cyclic $AKLM$ by P. (P and L might coincide if the

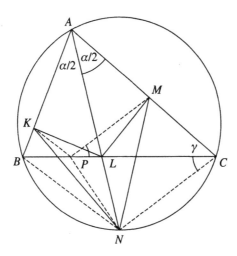

Figure 87/2.1

triangle is isosceles.) By the theorem of inscribed angles in this circle and also in the circumcircle of $\triangle ABC$

$$\frac{\angle A}{2} = \angle LAM = \angle LPM; \qquad \frac{\angle A}{2} = \angle BAN = \angle BCN = \angle PCN;$$

hence the segments MP and NC separated by the line PC making equal angles with the latter; therefore, they are parallel. Hence $\triangle MNC$ and $\triangle PNC$ have equal area: $[MNC] = [PNC]$. Replacing K and B with M and C, respectively, we get, similarly, that $[KNB] = [PNB]$. This is the end since the triangles PNC and PNB add up to the triangle BNC and thus $[MNC] + [KNB] = = [BNC]$, indeed.

Second solution. There are many ways to prove the claim using various well known metric relations of triangle geometry. The following is an option. We shall use the notations of the previous solution.

It is well known (c.f. the Note) that if h is the bisector of $\angle A$ then

$$AN = h = \frac{b+c}{2\cos\frac{A}{2}}.$$

Being symmetric with respect to the bisector of $\angle A$ $[AKNM] = \frac{1}{2}AN \cdot KM$.

The angle bisector theorem, on the other hand, yields $CL = \frac{ab}{b+c}$. Hence in the right triangle CML $ML = CL\sin C$; in $\triangle MKL$ we have $KM = = 2ML\sin\left(\frac{\pi - A}{2}\right) = 2ML\cos\frac{A}{2}$. Putting all these together:

$$[AKNM] = \frac{1}{2} \cdot \frac{b+c}{2\cos\frac{A}{2}} \cdot 2ML\cos\frac{A}{2} = \frac{1}{2}(b+c)\frac{ab}{b+c}\sin C = \frac{ab\sin C}{2} = [ABC],$$

the desired result.

Note. We prove now the result about the length of the chord bisecting $\angle A$. In the isosceles triangle BNC $BN = NC = \dfrac{a}{2\cos\frac{A}{2}}$. *Ptolemy's theorem* in the cyclic $ABNC$ yields

$$\frac{ab}{2\cos\frac{A}{2}} + \frac{ac}{2\cos\frac{A}{2}} = AN \cdot a,$$

and hence

$$AN = \frac{b+c}{2\cos\frac{A}{2}}.$$

1987/3. *Let* $x_1, x_2, \ldots, x_n$ *be real numbers satisfying*

(1) $$x_1^2 + x_2^2 + \ldots + x_n^2 = 1.$$

Prove that for every integer $k \geq 2$ *there are integers* a_i $(i = 1, 2, \ldots, n)$, *not all zero, such that* $|a_i| \leq k - 1$ *for all* i, *and*

(2) $$|a_1 x_1 + a_2 x_2 + \ldots + a_n x_n| \leq \frac{(k-1)\sqrt{n}}{k^n - 1}.$$

Solution. Applying the A.M-Q.M inequality for (1) yields

$$\frac{1}{\sqrt{n}} = \sqrt{\frac{x_1^2 + x_2^2 + \ldots + x_n^2}{n}} \geq \frac{|x_1| + |x_2| + \ldots + |x_n|}{n},$$

that is

$$|x_1| + |x_2| + \ldots + |x_n| \leq \sqrt{n}.$$

Consider now the $k^n - 1$ sequences of length n, none of them is constantly zero, that are composed from the numbers $(0, 1, 2, \ldots, k-1)$; let one of them be $(a_1, a_2, \ldots, a_n)$. Set now the sign of a_i in such a way that the product $a_i x_i$ is not negative $(i = 1, 2, \ldots, n)$. Then clearly

(3) $$a_1 x_1 + \ldots + a_n x_n = |a_1||x_1| + |a_2||x_2| + \ldots + |a_n||x_n| \leq$$
$$\leq (k-1)(|x_1| + |x_2| + \ldots + |x_n|) \leq (k-1)\sqrt{n}.$$

Split the interval $[0; (k-1)\sqrt{n}]$ into $k^n - 1$ equal parts; the length of each part is $\dfrac{(k-1)\sqrt{n}}{k^n - 1}$. If any one of the sums above is inside the first interval, then (2) clearly holds for this sum and we are done. If there is no such a sum then, by the pigeonhole principle, there are at least two of them in some of the remaining $k^n - 2$ intervals. If these sums are

$$b_1 x_1 + b_2 x_2 + \ldots + b_n x_n \quad \text{and} \quad c_1 x_1 + c_2 x_2 + \ldots + c_n x_n \quad (|b_i|, |c_i| \leq k - 1),$$

then, obviously, their difference cannot exceed the length of the interval:

$$|b_1 x_1 + b_2 x_2 + \ldots + b_n x_n - c_1 x_1 - c_2 x_2 - \ldots - c_n x_n| =$$
$$= |(b_1 - c_1)x_1 + (b_2 - c_2)x_2 + \ldots + (b_n - c_n)x_n| \leq \frac{(k-1)\sqrt{n}}{k^n - 1}.$$

Since b_i and c_i are of the same sign, $|b_i - c_i| \leq k - 1$. Now clearly $a_i = b_i - c_i$ are integers of the desired property:

$$|a_1 x_1 + a_2 x_2 + \ldots + a_n x_n| \leq \frac{(k-1)\sqrt{n}}{k^n - 1}.$$

Remark. (3) can be obtained immediately from (1) applying *Cauchy's inequality* ([22]) and then $a_i^2 \leq (k-1)^2$; indeed

$$a_1 x_2 + a_2 x_2 + \ldots + a_n x_n \leq \sqrt{a_1^2 + a_2^2 + \ldots + a_n^2} \sqrt{x_1^2 + x_2^2 + \ldots + x_n^2} \leq$$
$$\leq \sqrt{n(k-1)^2}\sqrt{1} = (k-1)\sqrt{n}.$$

1987/4. *Prove that there is no function f from the set of non-negative integers into itself such that*

(1) $$f(f(n)) = n + 1987$$

for all n.

Solution. Assume the contrary. Substituting $f(n)$ for n in (1):

(2) $$f(f(f(n))) = f(n) + 1987,$$

and plugging also the respective sides of (1) for n

(3) $$f(f(f(n))) = f(n + 1987).$$

Combining (2) and (3) yields

(4) $$f(n + 1987) = f(n) + 1987.$$

Let t be an arbitrary positive integer. We now prove by mathematical induction that

(5) $$f(n + 1987t) = f(n) + 1987t.$$

Fix the value of n. For $t = 1$ the claim is but (4). By the induction hypothesis

$$f(n + 1987(t-1)) = f(n) + 1987(t-1).$$

Substituting here $n + 1987$ for n (4) implies

$$f(n + 1987t) = f(n + 1987) + 1987(t-1) =$$
$$= f(n) + 1987 + 1987(t-1) = f(n) + 1987t,$$

which is (5), the proof is complete.

Let s be now an arbitrary non negative integer less than 1987 and consider the remainder r when $f(s)$ is divided by 1987.

$$f(s) = 1987k + r. \qquad (0 \le k \text{ and } 0 \le r \le 1986), leqno(6)$$

since, by condition, $f(s)$ is not negative. Hence by (1)

$$f(f(s)) = s + 1987,$$

and, from (6) and (5)

$$f(f(s)) = f(1987k + r) = f(r) + 1987k.$$

The last two results imply

(7) $$s + 1987 = f(r) + 1987k.$$

Since $s < 1987$

$$f(r) + 1987k < 2 \cdot 1987, \qquad f(r) < 1987(2 - k).$$

Given that $f(r) \ge 0$ there are two possible values of k: either $k = 1$ or $k = 0$. In the first case (6) and (7) yield

(8) $$f(s) = 1987 + r,$$

(9) $$f(r) = s,$$

and this overthrows $r = s$; indeed, plugging s for r in (8) and (9) forces $1987 = 0$, a contradiction.

In the other case $k = 0$. Combining again (6) and (7) implies

(10) $$f(s) = r,$$

(11) $$f(r) = 1987 + s;$$

and, like before, $r = s$ leads to a contradiction.

Making the two ends meet (9) and (10) together imply that when acting on the numbers

$$0, \ 1, \ 2, \ \ldots, \ 1986$$

f is arranging them into pairs (a, b) in such a way that either

$$f(a) = b \quad \text{and} \quad f(b) = a + 1987,$$

or $$f(b) = a \quad \text{and} \quad f(a) = b + 1987,$$

and, of course, the numbers in each pair are different. Now this is a contradiction since the number of elements of the set $0, 1, \ldots, 1986$ is odd.

Remark. The claim holds, of course, for any odd positive number instead of 1987.

1987/5. *Let n be an integer greater or equal to 3. Prove that there is a set of n points in the plane such that the distance between any two points is irrational and each set of 3 points determines a non-degenerate triangle with rational area.*

First solution. In the solutions we shall use a well known result about the area of polygons formed by lattice points i.e., points of integer coordinates: the area of such lattice polygons is a rational number. Consider now n *independent* lattice points. (The points of a set are independent if the are no three collinear one among them.) A simple induction argument shows that this can be done: indeed, any finite set of points determines but a finite set of lines so these lines cannot contain every lattice point; hence any finite set of independent points can be increased.

As it was noted, the respective areas of the triangles formed by these points is rational. Consider now the pairs from this set of points and denote the square of their distances by $d_1, d_2, \ldots, d_k$, respectively. Being calculated in the usual manner from integer coordinates these numbers are also whole numbers. Let p be a prime which is not dividing any one of these d_i and magnify the lattice by $\sqrt{p}$.

Consider now the image of our set of n independent points. Multiplied by p the area of each magnified triangle is still rational. Distances, on the other hand, are scaled up by $\sqrt{p}$ and thus the ith one magnifies into $\sqrt{pd_i}$. By the choice of p its index in the factorisation of pd_i is 1, pd_i is not a square and thus its square-root is irrational, indeed, which completes the proof.

Second solution. Pick n lattice points on the parabola $y = x^2$. The coordinates of these points are of the form (a, a^2) where a is integer.

Since any straight line meets the curve at two points, at most, there are no 3 collinear among the selected points and the area of the triangles formed by them is a rational number, as that of any lattice triangle,

The distance of the points $A(a, a^2)$, $B(b, b^2)$ is

$$AB = \sqrt{(a-b)^2 + (a^2-b^2)^2} = |a-b|\sqrt{1 + (a+b)^2}.$$

Since there is no positive square whose neighbour is also a square, $(a+b)^2 + 1$ is not a square, its square root is irrational; $a - b$ is a whole number different from zero and thus $|a-b|\sqrt{(a+b)^2 + 1}$ is also irrational: the pairwise distances in our set are irrational numbers.

Note. The general result about the area of a lattice polygon clearly follows from the special case of the triangle. Now any lattice triangle can be inscribed into a rectangle whose sides are lattice lines and thus its vertices are lattice points. Hence the area of the latter is an integer and the triangle itself can be obtained by removing right triangles of integral legs from the rectangle decreasing its area by the half of an integer each time: the area of our lattice triangle is indeed rational; it is, in fact, the half of a whole number.

This result also follows from the triangle area formula: if the (integral) coordinates of the vertices are (x_1, y_1), (x_2, y_2), (x_3, y_3), respectively then the area of the triangle computes as

$$t = \frac{1}{2} \begin{vmatrix} x_1 & y_1 & 1 \\ x_2 & y_2 & 1 \\ x_3 & y_3 & 1 \end{vmatrix} = \frac{1}{2}(x_1(y_2 - y_3) + x_2(y_3 - y_1) + x_3(y_1 - y_2)).$$

The celebrated result, known as *Pick's theorem* about the general lattice polygon states that its area is equal to

$$A = i - 1 + \frac{b}{2},$$

where i and b denote the number of lattice points inside and on the boundary of the polygon, respectively. This also yields, as a byproduct, that the area of lattice polygons is rational.

1987/6. *Let n be an integer greater or equal to 2. Prove that if $k^2 + k + n$ is prime for all integers such that $0 \le k \le \sqrt{\frac{n}{3}}$, then $k^2 + k + n$ is prime for all integers k such that $0 \le k \le n - 2$.*

Solution. Set $f(k) = k^2 + k + n$. What is to be proved, in fact, is the following: if $f(0), f(1), \ldots, f\left(\left[\sqrt{n/3}\right]\right)$ are all primes then so are the numbers $f(0), f(1), \ldots, f(n-2)$.

If the claim is false then there is a lowest non negative integer y such that $y \le n - 2$ and $f(y)$ is composite. (Since $n \ge 2$ $f(y) \ge 2$ so if not a prime it has to be composite.) By the choice of y for any k such that $0 \le k \le y - 1$ $f(k)$ is a prime number.

Denote the smallest prime divisor of the hence composite $f(y)$ by q. first we show that $q > 2y$. Assume the contrary and consider the differences

(1) $$f(y) - f(k) = y^2 + y + n - \left(k^2 + k + n\right) = (y - k)(y + k + 1)$$

as the integer k varies from 0 to $y - 1$. The array below displays the values of k and the factors, $y - k$ and $y + k + 1$, of the respective differences:

k:	0,	1,	2,	...,	$y - 2$	$y - 1$;
$y - k$:	y,	$y - 1$,	$y - 2$,	...,	2,	1;
$y + k + 1$:	$y + 1$,	$y + 2$,	$y + 3$,	...,	$2y - 1$,	$2y$.

One can check that the union of the ranges of the respective factors, $y - k$ and $y + k + 1$ contains every integer from 1 to $2y$ and thus, in particular, it contains q which, by assumption, is not exceeding $2y$. Hence there is a value of k such that $0 \le k \le y - 1$ and $(y - k)(y + k + 1) = f(y) - f(k)$ is a multiple of q. q, by its choice, divides $f(y)$, therefore it also divides $f(k)$. On the other hand, both $f(k)$ and q are prime numbers which leaves $f(k) = q$ as the only option.

By the choice of y, however

$$y - k \leq n - 2 < n + k + k^2 = f(k) = q,$$
$$y + k + 1 \leq n - 1 + k < n + k + k^2 = f(k) = q.$$

These inequalities imply that neither $y - k$ nor $y + k + 1$ is a multiple of q. Hence, as a prime, q cannot divide their product, $f(y) - f(k) = f(y) - q$ so it does not divide $f(y)$ contradicting its choice. The proof of $q > 2y$ is complete and observe also that since q is an integer, this slightly improves to $q \geq 2y + 1$.

We are almost at the end now. Since q is the lowest prime divisor of the composite $f(y)$, they are different, moreover

$$f(y) \geq q^2 \geq (2y + 1)^2 = 4y^2 + 4y + 1,$$
$$y^2 + y + n \geq 4y^2 + 4y + 1.$$

Rearranging we get $3y^2 + 3y + 1 - n \leq 0$. Satisfying this inequality, y cannot exceed the higher root of the polynomial on the *l.h.s.*:

$$y \leq -\frac{1}{2} + \sqrt{\frac{n}{3} - \frac{1}{12}}, \quad \text{and thus} \quad y < \sqrt{\frac{n}{3}}.$$

According to the condition, however, if y satisfies this inequality then $f(y)$ is a prime and thus $f(0), f(1), \ldots, f(n - 2)$ are all primes, a contradiction.

Remark. The curious way that certain prime values force further values of a polynomial to be prime makes this one a really intriguing problem. If $n = 41$, for example, then having checked that $k^2 + k + 41$ is prime for $k = 1, 2, 3$ the further 36 values, according to the problem, are also prime numbers: $61, 71, \ldots, 1061$. (This perplexing property of $x^2 + x + 41$ was, by the way, known to *L. Euler* in the 18th century.

Being so, it was straightforward to ask for those values of n which, indeed, have the property described in the problem. This has been undecided for a long period, to the extent, that even false assertions have been announced; in an 1939 paper $n = 72\,491$ was stated to have this property; this, however, has turned out to be false. A result of *H. M. Stark* from as late as 1967 implies that there are seven such values of n altogether; they are

$$n = 1, 2, 3, 5, 11, 17, 41.$$

There are certain algebraic number theoretical investigations of complex numbers in the background of the problem. The reader can find further information about the topic in *E. Gyarmati: A note on my paper: "Unique prime factorization in imaginary quadratic number fields"* (Annales Univ. Sci. Budapest., Sectio Mathematica, **26** (1983), 195–196.)

1988.

1988/1. *Consider two coplanar circles of radii R and r $(R > r)$ with the same centre. Let P be a fixed point on the smaller circle and B a variable point on the larger circle. The line BP meets the larger circle again at C. The perpendicular l to BP at P meets the smaller circle again at A (if l is tangent to the circle at P then $A = P$).*

(i) *Find the set of values of $BC^2 + CA^2 + AB^2$.*

(ii) *Find the locus of the midpoint of AB.*

Solution. Denote the common centre of the two circles by O. The smaller circle divides the segment BC into parts x, y and x again. The chord of the line l inside the smaller circle is z. Both y and z might reduce to zero if BC, or l are tangent to the smaller circle (*Figure 1988/1.1*).

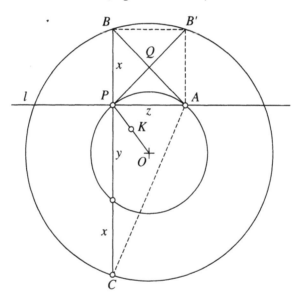

Figure 88/1.1

Applying *Pythagoras' Theorem* on the perhaps degenerate triangles APC and APB:

$$AB^2 + BC^2 + CA^2 = x^2 + z^2 + (2x + y)^2 + (x + y)^2 + z^2 = 6(x + y)x + 2(y^2 + z^2).$$

As perpendicular chords in the smaller circle, y and z are the legs of a right triangle of hypotenuse $2r$ and thus $y^2 + z^2 = 4r^2$; furthermore, the power of P with respect to the larger circle is $(x + y)x = R^2 - r^2$ and hence

$$AB^2 + BC^2 + CA^2 = 6(R^2 - r^2) + 8r^2 = 6R^2 + 2r^2$$

is constant; the first part of the problem is hence settled. We note here that B and C can be swapped in the argument but this, of course, has no effect on the result.

Moving on to the second part denote the mirror image of B through the perpendicular bisector of AP by B' (if $P = A$ then, of course, $B = B'$). The respective points clearly form a rectangle $PAB'B$ whose centre, Q, is the midpoint of AB. Thus the points Q in question are produced by reducing the larger circle from P by half. K, the centre of this circle k is the midpoint of PO and its radius is $\dfrac{R}{2}$.

Each point of k belongs to the locus: indeed, if Q is on the diameter PO then $P = A$ and thus B and C are the respective endpoints of the diameter PO. If Q is not on PO then enlarging it from P by a scale factor 2 yields a point B' on the larger circle. Now clearly $PB' > R - r$ and thus the *Thales*-circle of diameter PB' meets the larger circle once more at some point B. For the corresponding A the point Q is clearly the midpoint of AB that is Q belongs to the locus, indeed.

The locus of the midpoint of BC is the circle of diameter R about the midpoint of OP.

1988/2. *Let n be a positive integer and let A_1, A_2, ..., A_{2n+1} be subsets of a set B. Suppose that*

a) *each A_i has exactly $2n$ elements,*

b) *each $A_i \cap A_j$ ($1 \leq i < j \leq 2n+1$) contains exactly one element, and*

c) *every element of B belongs to at least two of the A_i.*

For which values of n can one assign to every element of B one of the numbers 0 and 1 in such a way that each A_i has 0 assigned to exactly n of its elements?

First solution. We start by showing that each element of B belongs to exactly two A_i.

The problem is worth rephrasing in graph terminology. Consider a graph G of $s + 2n + 1$ vertices. Assign each element b_1, b_2, ..., b_s of B and also the subsets A_i, $i = 1, 2, \ldots, 2n+1$ to the vertices of G, labelling each vertex by corresponding symbol. Vertex A_i is connected to vertex b_j if and only if b_j is in A_i; these are the edges in the graph: G is a so called *bipartite* graph. The conditions when also rephrased become:

a) the degree of each vertex A_i is $2n$,

b) for any two vertices A_i and A_j there is exactly one vertex b_k connected to both of them,

c) the degree of any vertex b_j is at least 2 (*Figure 1988/2.1*).

We are to prove that the degree of every element b_j of B is exactly 2. Let $b \in B$ arbitrary. Its degree, by c), is at least 2. Assume, to the contrary, that there are three edges incident to b connecting it to the vertices A_1, A_2 and A_3, for

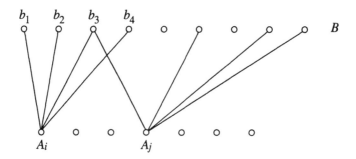

Figure 88/2.1

example. From A_1 there are further $2n - 1$ edges, by a), connecting it to b_1, b_2, ..., b_{2n-1} and, similarly, A_2 and A_3 are adjacent to the vertices b_{2n}, ..., b_{4n-2} and b_{4n-1}, ..., b_{6n-3}, respectively.

By b), none of these, altogether $6n - 3$ vertices are connected to any one of A_1, A_2 and A_3. Indeed, if, for example, b_{2n} would be connected to A_3, for example, then the pair A_2 and A_3 would, in fact, be adjacent to two elements of B, namely b and b_{2n} violating b). The $6n - 3$ neighbours of A_1, A_2 and A_3 listed above are thus all different.

Hence, by c), there are at least $6n - 3$ edges connecting the above listed b_1, b_2, ..., b_{6n-3} to the remaining A-vertices, A_4, A_5, ..., A_{2n+1}. There is, therefore, some A_i ($4 \le i \le 2n + 1$), adjacent to at least four of the b-s listed above since $3(2n - 2) < 6n - 3$. Without loss of generality this A-vertex can be assumed to be A_4. There are four edges connecting A_4 to the three sets of vertices $\{b_1, \ldots, b_{2n-1}\}$, $\{b_{2n}, \ldots, b_{4n-2}\}$ and $\{b_{4n-1}, \ldots, b_{6n-3}\}$ so two of these edges enter the same set: they arrive, for example, to the vertices b_1 and b_2, for example. Now this is a contradiction, since A_1 and A_4 are thus connected to two b-s which is prohibited by b).

Hence the degree of each b_j is exactly 2, indeed.

Assigning now 0 or 1 to some b_j in the graph will be carried out by colouring this b_j red or green, respectively, and, at the same time, by colouring also the edges starting from this b_j with the very same colour. Hence, if the assignment of the problem is feasible then the edges of G are all coloured and their colour is the same as that of their b-endpoint. Having thus coloured a proper assignment there are n red and n green edges leaving each vertex A_i therefore there are, altogether, $(2n + 1)n$ red edges in G.

Any red edge is between some A_i and a (red, of course) b_k; the degree of the latter is 2 and thus there is another (red, of course) edge connecting it to some A_j. This simple observation implies that the red edges can be (uniquely) divided into pairs, their total number, $(2n + 1)n$ is hence even and thus n itself is also even. This is necessary for the required 0-1 assignment to exist.

We show now, representing the sets in question, that the condition for n to be even is also sufficient. For arbitrary even value of n mark, on a circle, the vertices $A_1, A_2, \ldots, A_{2n+1}$ of a regular $2n+1$-gon. Connect each vertex to the neighbouring $\frac{n}{2}$ vertices to its left and also to its right. Colour these segments red and all the remaining diagonals green. Under this colouring there are clearly n red and n green edges leaving each vertex. The respective sets, together with the required assignment are now the following:

set B consists of the — coloured — edges, the red ones are labelled by 0 and the green ones by 1. The subsets A_i are the edges leaving the vertex A_i.

a), b) and c) do obviously hold for this construction: there are $2n$ elements of each A_i; the edge connecting the corresponding vertices is the single common element of A_i and A_j; finally any element of B is contained by exactly two A_i, these are the ones labelled by its endpoints.

Second solution. In what follows, a general approach to the first part of the problem, is outlined.

Let B have s elements, $b_1, b_2, \ldots, b_s$. The set B together with its $2n+1$ subsets $A_1, A_2, \ldots, A_{2n+1}$ can be displayed in a $(2n+1)$ by s array (matrix) whose entries are 0 and 1, the so called *incidence matrix* of the graph G of the first solution. The columns — there are s of them — are to represent the elements of B. Write 1 in the respective entry if the very element b corresponding to its column belongs the subset corresponding to its row and 0 otherwise: the rows are hence for the $2n+1$ subsets.

	b_1	b_2	b_3		b_s
A_1	1	0	1	$\ldots$	1
A_2	1	1	0	$\ldots$	0
A_3	0	0	1	$\ldots$	0
				$\vdots$	
A_{2n+1}	0	1	0	$\ldots$	1

The number of 1 entries, by the condition, is $2n$ in each row. Thus, if M is the total number of 1-s in the array then $M = 2n(2n+1)$. Denote, additionally, the number of 1-s in the respective columns by $c_1, c_2, \ldots, c_s$, respectively. Clearly

$$M = c_1 + c_2 + \ldots + c_s.$$

The ith row of the array, for brevity, shall be considered an s-dimensional vector $\mathbf{a}_i$ whose components are the 0 and 1 entries of the respective rows. Thus, for example

$$\mathbf{a}_1(1, 0, 1, \ldots, 1).$$

(See [39].) The conditions of the problem are then

a) $\mathbf{a}_i^2 = 2n$ $(i = 1, 2, \ldots, 2n+1)$;

b) $\mathbf{a}_i \mathbf{a}_j = 1$, $(i \neq j)$;

c) $c_i \geq 2$, $(i = 1, 2, \ldots, s)$, and thus $M \geq 2s$ that is $s \leq n(2n+1)$. Let $\mathbf{a}$ be the sum of the vectors $\mathbf{a}_i$. Since its components are $(c_1, c_2, \ldots, c_s)$, the square of the vector $\mathbf{a}$ computes as

(1)
$$(\mathbf{a})^2 = c_1^2 + c_2^2 + \ldots + c_s^2.$$

On the other hand, with a bit of algebra

(2)
$$(\mathbf{a})^2 = \sum_1^{2n+1} \mathbf{a}_i^2 + 2\sum_{i<j} \mathbf{a}_i \mathbf{a}_j = (2n+1)2n + 2\binom{2n+1}{2} = 4n(2n+1).$$

The A.M.–Q.M. inequality provides a lower bound for the r.h.s. of (1) and thus

(3)
$$(\mathbf{a})^2 \geq s\left(\frac{\sum_1^s c_i}{s}\right)^2 = \frac{M^2}{s} = \frac{1}{s}\cdot 4n^2(2n+1)^2.$$

Comparing (2) and (3) yields
$$s \geq n(2n+1).$$
This when combined with the inequality c) forces $s = n(2n+1)$; there is equality in (3) and hence the numbers c_i are all equal. Thus $c_i = 2$ for every i that is each element b is contained in exactly two subsets, indeed.

Note. Set B with its subsets A_i as they are given forms a so called *block-design*. Their research is an important task of combinatorics ([31]).

1988/3. *A function f is defined on the positive integers by*
(1) $$f(1) = 1, \quad f(3) = 3,$$
(2) $$f(2n) = n,$$
(3) $$f(4n+1) = 2f(2n+1) - f(n),$$
(4) $$f(4n+3) = 3f(2n+1) - 2f(n)$$
for all positive integers n.

Determine the number of positive integers n, less than or equal to 1988, for which $f(n) = n$.

Solution. Note, first of all, that formulas (1)–(4) determine function f unambiguously, since the numbers of the form $2n, 4n+1$ and $4n+3$ exhaust the set of integers. Hence it is enough to provide a function f whatsoever satisfying conditions (1)–(4).

Check a few values of f:

n	1	2	3	4	5	6	7	8	9
$f(n)$	1	1	3	1	5	3	7	1	9

Far from a pattern but as for the values satisfying $f(n) = n$ they seem to show up next to the powers of 2. This might suggest to rewrite the rows in base 2:

n_{10}	1	2	3	4	5	6	7	8	9
n_2	1	10	11	100	101	110	111	1000	1001
$f(n)_2$	1	01	11	001	101	011	111	0001	1001.

Now it is hard to resist the suspicion that f, when in action, inverts the order of digits in the binary representation of n. We shall, in fact, prove this by induction on k, the number of binary digits of n.

Note, first of all, that the last binary digits of the numbers $2n$, $2n+1$, $4n$, $4n+1$, $4n+3$ are respectively

$$\begin{array}{ccccc} 2n & 2n+1 & 4n & 4n+1 & 4n+3 \\ 0 & 1 & 00 & 01 & 11 \end{array}$$

In what follows binary strings are going to be overlined. If $k=1$ or 2 then the claim has already been verified above. Let $k>2$ and assume that it holds for any number whose binary form consists of less than k digits, i. e., if $n = \overline{a_{k-1}a_{k-2}\dots a_1}$ ($a_{k-1}=1$); then

(5) $$f(n)=\overline{a_1 a_2 \dots a_{k-1}}.$$

1. Set now $m=2n=\overline{a_1 a_2 \dots a_{k-1} 0}$. Then clearly

$$f(m)=f(2n)=f(n)=\overline{a_{k-1}a_{k-2}\dots a_1}=\overline{0 a_{k-1}a_{k-2}\dots a_1},$$

(5) holds for k-digit even numbers. The odd case is checked in two parts.

2. Let $m=4n+1=\overline{a_1 a_2 \dots a_{k-2} 01}$, where n has now $k-2$ binary digits ($k>2$), $n=\overline{a_1 a_2 \dots a_{k-2}}$. Hence $2n+1=\overline{a_1 a_2 \dots a_{k-2} 1}$, and thus, by (3)

$$f(m)=2f(2n+1)-f(n)=f(2n+1)+(f(2n+1)-f(n))=$$
$$=\overline{1 a_{k-2}\dots a_2 a_1}+\overline{1 a_{k-2}\dots a_2 a_1}-\overline{a_{k-2}\dots a_2 a_1}=$$
$$=\overline{1 a_{k-2}\dots a_2 a_1}+\overline{10\dots 0}=\overline{10 a_{k-2}\dots a_2 a_1},$$

so (5) holds for k-digit numbers of the form $4n+1$ as well.

3. Let $m=4n+3=\overline{a_1 a_2 \dots a_{k-2} 11}$, where n, like before, is equal to $\overline{a_1 a_2 \dots a_{k-2}}$. Using (4) again

$$f(m)=3f(2n+1)-2f(n)=f(2n+1)+2(f(2n+1)-f(n))=$$
$$=\overline{1 a_{k-2}\dots a_1}+2\left(\overline{1 a_{k-2}\dots a_1}-\overline{a_{k-2}\dots a_2 a_1}\right)=$$
$$=\overline{1 a_{k-2}\dots a_1}+\overline{10\dots a_1}=\overline{11 a_{k-2}\dots a_1},$$

that is (5) holds for k-digit numbers of the form $4n+3$ and this one was left to be checked for the induction to be complete.

Having understood the way f works it is obvious that $f(n)=n$ for those numbers whose binary form is symmetric that is reversing the order of their digits yields the very same string. These are the so called *palindromes*, like 10111101, for example.

The leading digit of a $2n$ long binary palindrome is 1 and there are 2^{n-1} ways to set the next $n-1$ digits; hence there are this many $2n$ digit binary palindromes, altogether. The number of $2n+1$ long binary palindromes is clearly twice this much since each of them can be obtained from a unique $2n$ long one by inserting a single 0 or 1 into the middle.

As $2^{10} < 1988 < 2^{11}$, the binary form of 1988 has 11 digits. The previous observations hence imply that the number of at most 11 long binary palindromes is the sum

$$2^0 + 2^0 + 2^1 + 2^1 + 2^2 + 2^2 + 2^3 + 2^3 + 2^4 + 2^4 + 2^5 = 94.$$

We are not finished yet. $1988 = \overline{11111000100}$ and thus there are two 11-digit binary palindromes above 1988: $\overline{11111011111}$ and $\overline{11111111111}$. Hence the correct answer is 92.

1988/4. *Show that the set of real numbers x which satisfy the inequality*

(1)
$$\sum_{k=1}^{70} \frac{k}{x-k} \geq \frac{5}{4}$$

is a union of disjoint intervals, the sum of whose lengths is 1988.

Solution. Denote the function on the *l.h.s.* of (1) by $f(x)$; its domain is the set obtained by removing the numbers 1, 2, ..., 70 from the set of reals; it is the union of the open intervals

$$]-\infty, 1[, \quad]1, 2[, \quad]2, 3[, \quad \ldots, \quad]69, 70[, \quad]70, +\infty[.$$

The terms of the sum defining f are all continuous and decreasing inside these intervals and so is their sum, f. In the open interval $]-\infty, 1[$ f is negative; at the values k ($k = 1, 2, \ldots, 70$) the terms $f_i(x) = \dfrac{i}{x-i}$ ($i \neq k$) are continuous and bounded; the limit of f_k, as x tends to k from the left and from the right is $-\infty$ and $+\infty$, respectively, so this holds for f, too. This means that as a continuous function in the interval $(k, k+1)$ ($1 \leq k < 70$) varying decreasingly from $+\infty$ at the left endpoint to $-\infty$ at the right one, f admits every real value in this interval. As for the last interval, $]70, +\infty[$ here f is positive, its limit at the lower endpoint from the right is $+\infty$, and at $+\infty$ it is 0. Being so f admits every positive value in the interval $]70, +\infty[$. (To make life simpler on *Figure 1988/4.1* we have sketched the 3 term sum f when k varies from 1 to 3.)

The diagram also shows that, apart from the first one, $[-\infty, 1[$, f admits the value $\dfrac{5}{4}$ in every interval, each of which, hence, contains a subinterval where $f \geq \dfrac{5}{4}$ (f is decreasing).

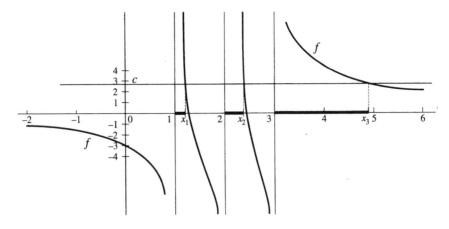

Figure 88/4.1

If f admits the value $\frac{5}{4}$ at x_1, x_2, $\ldots$, x_{70} ($k < x_k < k+1$, if $k \leq 69$ and $x_{70} > 70$) then $f \geq \frac{5}{4}$ in the disjoint intervals

$$]1, x_1], \;]2, x_2], \;\ldots, \;]69, x_{69}], \;]70, x_{70}].$$

Their total length is:

(2)

$$H = (x_1 - 1) + (x_2 - 2) + \ldots + (x_{70} - 70) = (x_1 + x_2 + \ldots + x_{70}) - (1 + 2 + \ldots + 70).$$

The task remaining is to calculate the sum $x_1 + x_2 + \ldots + x_{70}$. To make it simpler switch to the function $g(x) = f(x) - \frac{5}{4}$. The zeros of this g are clearly the previous x_i values. It is a first degree rational function, its denominator is $(x - 1)(x - 2) \ldots (x - 70)$, while its numerator is

$$1 \cdot (x - 2) \cdot (x - 3) \ldots (x - 70) + 2 \cdot (x - 1)(x - 3) \ldots (x - 70) + \ldots$$

$$\ldots + 70 \cdot (x - 1)(x - 2) \ldots (x - 69) - \frac{5}{4} \cdot (x - 1)(x - 2) \ldots (x - 70).$$

The leading coefficient of this 70-degree polynomial is $-\frac{5}{4}$ and thus the coefficient of the 69th degree term can be computed by the corresponding *Viete-formula* as

$$x_1 + x_2 + \ldots + x_{70} = \frac{4}{5}\left(1 + 2 + \ldots + 70 + \frac{5}{4}(1 + 2 + \ldots + 70)\right) =$$

$$= \frac{9}{5}(1 + 2 + \ldots + 70).$$

Hence, by (2), the total length of the intervals is

$$\frac{9}{5}(1 + 2 + \ldots + 70) - (1 + 2 + \ldots + 70) = \frac{4}{5} \cdot \frac{70}{2} \cdot 71 = 1988, \quad \text{indeed.}$$

1988/5. *ABC is a triangle right-angled at A, and D is the foot of the alti-tude from A. The straight line joining the incentres of the triangles ABD, ACD intersects the sides AB, AC at the points K, L respectively. S and T denote the areas of the triangles ABC and AKL respectively. Show that* $S \geq 2T$.

First solution. $AB \leq AC$ can clearly be assumed. Let the bisector f_1 of $\angle BAD$ and BD intersect at X and the bisector f_2 of $\angle CAD$ and CD at Y (*Figure 1988/5.1*). Reflect $\triangle ADX$ through f_1 obtaining $\triangle AD'X$. This reflec-

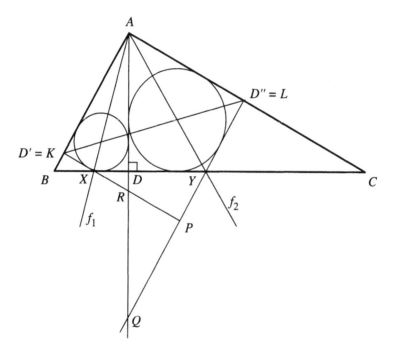

Figure 88/5.1

tion keeps the incircle of ABD fixed and thus this circle is also inscribed to the kite $AD'XD$. A similar procedure — reflection through f_2 — yields the cyclic kite $AD''YD$ whose incircle is the same as that of ACD. Denote, finally, the intersection of $D'X$ and $D''Y$ by P.

The quadrilateral $AD'PD''$ is a square since there are right angles at A, D' and D'' and, additionally, $AD = AD' = AD''$ because of the reflection. The diagonal $D'D''$ bisects the angles of the square at D' and D'' so it is passing through the centres of the two incircles. Therefore $D' \equiv K$ and $D'' \equiv L$, the right triangle AKL is isosceles, it is the half of the square $AD'PD''$.

Reflect now B through f_1 to R and C through f_2 to Q; these mirror images are clearly incident to the altitude AD and, by $AB \leq AC$, we have $AR \leq AQ$.

Using the equality of areas implied by the respective reflections:

$$S = [ABD] + [ABC] + [ACD] = [ARD'] + [AQD''] \geq [AD'PD''] =$$
$$= 2[AD'D''] = 2[AKL] = 2T,$$

indeed.

Second solution. Denote the incentres of ABD and ACD by O_1 and O_2, respectively (*Figure 1988/5.2*). The similarity properties of right triangles imply that the composition of the enlargement by $\frac{c}{b}$ and the rotation through $90°$, both about D is mapping ADC to BDA and O_2 to O_1. Hence $DO_1 : DO_2 = c : b$ and thus $\triangle DO_1O_2$ and $\triangle ABC$ are similar because the ratios of two corresponding sides are equal and these sides now make right angles, respectively. This similarity implies that $\angle DO_2O_1 = \angle C$. Hence the quadrilateral DO_2LC is cyclic and its exterior angle at L is equal to $\angle O_2DC = 45°$. Since $\angle ALK = 45°$ in the right triangle ALK, it is isosceles, $AL = AK$.

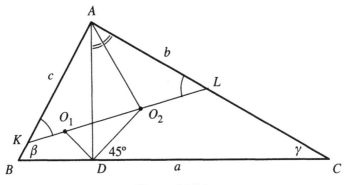

Figure 88/5.2

$\triangle ALO_2$ and $\triangle ADO_2$ are congruent because they have a common side, their respective angles at A are equal, finally $\angle ADO_2 = \angle ALO_2 = 45°$. Therefore $AL = AD$. Hence

$$T = [ALK] = \frac{AD^2}{2}.$$

Now $a \cdot AD = bc$ yields $AD = \dfrac{bc}{a}$ and since $b^2 + c^2 \geq 2bc$,

$$T = \frac{b^2c^2}{2a^2} = \frac{b^2c^2}{2(b^2 + c^2)} \leq \frac{bc}{4} = \frac{S}{2},$$
$$S \geq 2T.$$

Third solution. Denote the incentres of ADB and ADC by O_1 and O_2, the inradii by r_1 and r_2 respectively; also the touching points of the respective incircles on AB and AD by E and E' and on AC and AD by F and F' (*Figure*

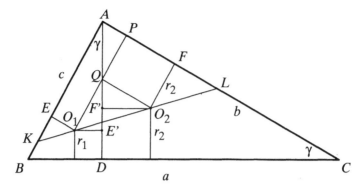

Figure 88/5.3

1988/5.3). The foot of the perpendicular from O_2 to O_1P is denoted by Q and the altitude AD by m.

We shall make use of the equality of the tangent segments to a circle from an external point. Express, in terms of m, r_1 and r_2, the distances O_1Q and O_2Q:

$$O_1Q = O_1P - QP = AE - O_2F = AE' - r_2 = m - r_1 - r_2,$$
$$O_2Q = FP = AF - AP = AF' - r_1 = m - r_2 - r_1.$$

This calculation shows that the right triangle O_1QO_2 is isosceles and thus, having their respective sides parallel, triangles KAL and O_2FL are also isosceles right triangles. Hence $FL = r_2$ and thus

$$AL = AF + r_2 = AF' + r_2 = m.$$

Since $m = c\cos C = b\sin C$

$$T = \frac{m^2}{2} = \frac{bc\sin C\cos C}{2} = \frac{bc}{4}\sin 2C = \frac{S}{2}\sin 2C,$$

so

$$T \le \frac{S}{2}, \qquad S \ge 2T,$$

and we are done.

Remark. Every solution shows that equality $S = 2T$ holds if and only if the right triangle ABC is isosceles.

1988/6. *Let a and b be positive integers such that $ab+1$ divides a^2+b^2. Show that*

(1)
$$\frac{a^2+b^2}{ab+1}$$

is the square of an integer.

Solution. Denote the value of the ratio (1) by q. The claim can be rephrased as follows: if there is a positive integer q such that the equation

(2) $a^2 - qab + b^2 - q = 0$

is satisfied by some pair (a, b) of positive integers then q is a square.

Suppose the contrary by assuming that (2) has a solution and q is not a square. Consider the quadratic

(3) $$x^2 - qxy + y^2 - q = 0 \qquad (q > 0 \text{ integer}).$$

Plugging numbers of opposite sign for x and y clearly $-qxy > q$, the l.h.s. of (3) is positive so equation (3) does not have roots — not even reals — of opposite sign. Even if one of the solutions, say x, for example, is zero then this would force $q = y^2$ but exactly this was assumed to be false. Since any solution (x, y) implies also $(-x, -y)$ to be a solution of (3), from now on we can restrict ourselves to the strictly positive integral solutions of (3). By condition this set is not empty.

Consider now an element of the set of positive integral solutions of (3) for which the sum of the square of the components is minimal; denote one of these solutions of (3) by (A, B); here $0 < B \le A$ can clearly be assumed. Hence

(4) $$A^2 - qAB + B^2 - q = 0$$

and there is no positive integral solution whose square sum is smaller than $A^2 + B^2$.

Now (4) also says that A is a positive integral solution of the quadratic

(5) $$x^2 - qBx + B^2 - q = 0.$$

Consider now the other solution A' of (5). By the previous observation A' is also positive. Writing down the formulas of *Viete*

(6) $A + A' = qB,$ (7) $AA' = B^2 - q.$

(6) implies the A' is also an integer, since A and qB are integers. (7) yields $AA' < B^2$ and this, together with $A \ge B$, shows that $A' < B$ and thus

$$A'^2 + B^2 < 2B^2 \le A^2 + B^2.$$

Now this is a contradiction. Indeed, the pair (A', B) is also a positive integral solution of (3), assuming that q is not a square we have managed to produce a solution of lower square sum than the certainly existing minimum. The proof is thus finished.

Remarks. 1. It is natural to ask if there exist positive numbers (a, b) of the given property at all. One can show that there are, in fact, infinitely many of them and the following recurrence lists them all:

$$x_0 = 0, \quad x_1 = a, \quad x_{n+1} = a^2 x_n - x_{n-1}, \qquad \text{where } a \text{ is a positive integer, } n \ge 1.$$

The solutions of (5) are the pairs (x_1, x_2), (x_2, x_3), $\ldots$, (x_{n-1}, x_n), $\ldots$; the value of q is a^2.

2. The method of the solution is called *infinite descent* in arithmetic. It is able to prove that a certain diophantine equation has no solution by choosing a particular hypothetical solution of some minimal positive integral measure and and then proceeding to another solution whose measure is definitely smaller. It was invented by *P. Fermat* in the 17th century.

1989.

1989/1. *Prove that the set* $\{1, 2, \ldots, 1989\}$ *can be expressed as the disjoint union of subsets* A_1, A_2, $\ldots$, A_{117} *in such a way that each* A_i *contains* 17 *elements and the sum of the elements in each* A_i *is the same.*

First solution. The subsets A_i will be formed of 7 pairs and one triple each, in such a way that the sum of the pairs and also that of the respective triples is the same in every subset. In the first step the altogether $117 \cdot 7 = 819$ pairs will be prepared and the 117 triples afterwards.

The first 351 integers are set aside for the triples and the pairs will be formed from the integers 352, 353, $\ldots$, 1989 in a straightforward manner:

$$(352, 1989)$$
$$(353, 1988)$$

(1)

$$\vdots$$

$$(1170, 1171).$$

These 819 pairs contain every number from 352 to 1989 and the sum of each pair is 2341, the same.

With a bit more care the remaining numbers can also be arranged into 117 triples, with the same sum each:

$(1, 176, 351)$	$(60, 118, 350)$
$(2, 177, 349)$	$(61, 119, 348)$
$(3, 178, 347)$	$(62, 120, 346)$

(2)

$\vdots$	$\vdots$
$(58, 233, 237)$	$(116, 174, 238)$
$(59, 234, 235)$	$(117, 175, 236).$

You can check that every integer, from 1 to 351 appears but once and the sum of each triple is 528.

The subsets A_i now can be assembled one by one as the union of 7 pairs from list (1) and a triple from list (2).

Second solution. As a more structural approach arrange the first 1989 integers in a 117 by 17 array in their natural order:

1	2	3	$\ldots$	17
18	19	20	$\ldots$	34

(1)

$$\vdots$$

1973	1974	1975	$\ldots$	1989

The task is now to rearrange the entries in such a way that the sum of the elements is the same in each row. If we succeed then the 117 rows will form the required subsets. The numbers in the jth column are clearly j, $j+1\cdot 17$, $j+2\cdot$ $\cdot 17$, ..., $j+116\cdot 117$. Denoting the i-th one by $j+a_{ij}\cdot 17$ $(0\le a_{ij}\le 116)$ the sum of the ith row is

$$s_i=\sum_{j=1}^{17}(j+a_{ij}\cdot 17)=\sum_{j=1}^{17}j+17\cdot\sum_{j=1}^{17}a_{ij}.$$

Since $\sum j$ is constant it is enough to calibrate the numbers a_{ij} in such a way that the sums $\sum a_{ij}$ are equal in the respective rows.

This can obviously be done with any even number of columns by simply reverting the order of numbers in every other column. We have to be careful, however, since the number of columns is now odd: the last extra column ruins balance. We need the last three of them to maintain equilibrium, as it is shown below.

(2)

$j=$	1	2	3	4	... ...	14	15	16	17
$i=$									
1	0	116	0	116		116	0	59	115
2	1	115	1	115		115	1	60	113
⋮	⋮	⋮				⋮	⋮	⋮	⋮
58	57	59	57	59		59	57	116	1
59	58	58	58	58		58	58	0	116
60	59	57	59	57		57	59	1	114
⋮	⋮	⋮				⋮	⋮	⋮	⋮
116	115	1	115	1		1	115	57	2
117	116	0	116	0		0	116	58	0

Array (2) clearly shows that preparing the subsets A_i as it is described above each number, from 1 to 1989 appears but once. In fact, the columns of array (1) are permuted according to the numbers in array (2).

1989/2. *In an acute-angled triangle ABC, the internal bisector of angle A meets the circumcircle again at A_1. Points B_1 and C_1 are defined similarly. Let A_0 be the point of intersection of the line AA_1 with the external bisectors of angles B and C. Points B_0 and C_0 are defined similarly. Prove that the area* *of the triangle $A_0B_0C_0$ is twice the area of the hexagon $AC_1BA_1CB_1$ and at least four times the area of the triangle ABC.*

Solution. The points A_0, B_0 and C_0 are clearly the excentres of the triangle. The bisectors of the respective interior and the exterior angles are perpendicular, AA_0, BB_0 and CC_0 are the altitudes in $\triangle A_0B_0C_0$, the triangle ABC is the *pedal* triangle of $\triangle A_0B_0C_0$. Therefore k, its circumcircle is the *nine points circle* of the same $\triangle A_0B_0C_0$. Moreover, the angle bisectors of ABC meet at K, the orthocentre of $\triangle A_0B_0C_0$ (*Figure 1989/2.1*).

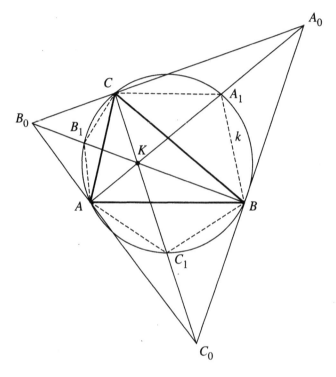

Figure 89/2.1

Since the nine points circle bisects the respective distances between the orthocentre and the vertices, A_1, B_1 and C_1 are the midpoints of KA_0, KB_0 and KC_0, respectively. According to $KA_1 = A_1A_0$, for example, $[KBA_1] = [A_1BA_0]$ because they have an equal side and the corresponding altitudes from B are identical. Similarly, $[KCA_1] = [A_1CA_0]$ but then also $[KBA_1C] = [BA_1CA_0]$. Repeating once more: $[KCB_1A] = [CB_1AB_0]$ and $[KAC_1B] = [AC_1BC_0]$. Adding the respective sides of these last three equalities the *l.h.s.* is the area of the hexagon $AC_1BA_1CB_1$ while the *r.h.s.* is the area of the region making the hexagon up to $\triangle A_0B_0C_0$ which proves the first proposition.

For the second one it is enough to show that the area of the hexagon is at least twice of $[ABC]$ since, as we have just proved, $[A_0B_0C_0]$ is the double of the hexagon's area.

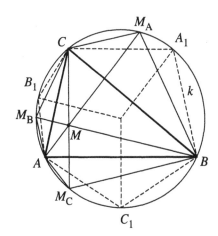

For the proof let us note first that among the triangles inscribed the minor arc AB of the circle k it is $\triangle ABC_1$ that has the greatest area because this one has the greatest altitude to AB (*Figure 1989/2.2*). Denote the orthocentre of ABC by M and its mirror image through AB by M_C. It is well known that M_C is lying on the circle k. As it was noted above

$$[AM_CB] \leq [AC_1B]$$

and equality holds if and only if $AC = = BC$. Similarly, $[BM_AC] \leq [BA_1C]$ and $[CM_BA] \leq [CB_1A]$. Being acute, $\triangle ABC$ contains M in its interior and thus the following decomposition is valid:

Figure 89/2.2

$$[ABC] = [AM_CB] + [BM_AC] + [CM_BA],$$
$$[AC_1BA_1CB_1] \geq [AM_CBM_ACM_B] = 2[ABC],$$

the proof is complete.

Note. There are several ways to finish the proof, the second proposition, for example, can be rephrased as

the area of an acute triangle is at least four times the area of its pedal triangle.

To prove this denote the sides of the triangle by a, b and c, respectively. If its circumradius is R then, as it is well known, the sides are $a = 2R \cdot \sin A$, $b = 2R \cdot \sin B$. The sides of the pedal triangle, on the other hand, are $a \cos A$, $b \cos B$ and $c \cos C$ and its angles are $180° - 2A$, $180° - 2B$ and $180° - 2C$, respectively. Its area is

$$t = \frac{ab \sin C}{2} = 2R^2 \sin A \cdot \sin B \cdot \sin C;$$

and that of the pedal triangle is

$$t' = \frac{a \cos A \cdot b \cos B \cdot \sin(180° - 2C)}{2} =$$
$$= 4R^2 \sin A \cdot \sin B \cdot \sin C \cdot \cos A \cdot \cos B \cdot \cos C.$$

The pending inequality $4t' \leq t$ hence becomes

$$16R^2 \sin A \cdot \sin B \cdot \sin C \cdot \cos A \cdot \cos B \cdot \cos C \leq 2R^2 \sin A \cdot \sin B \cdot \sin C.$$

This simplifies to the well known

$$\cos A \cdot \cos B \cdot \cos C \leq \frac{1}{8}.$$

This reformulation of the claim now can be smashed if invoking the generalization of the so called *Simson line theorem*. This big gun states that if the area of a triangle is a, its circumradius is R and the distance of its incentre from a given point P of the plane is d then the feet of the perpendiculars from P to the respective sides form a triangle whose signed area, α is equal to

$$\alpha = \frac{a(R^2 - d^2)}{4R^2} = a\left(\frac{1}{4} - \frac{d^2}{4R^2}\right).$$

Set now P as the orthocentre of the triangle. Then $\alpha = a'$ is the area of the pedal triangle and thus

$$t' \le \frac{t}{4}, \qquad 4t' \le t, \quad \text{indeed.}$$

Equality holds here if $d = 0$, that is the orthocentre and the circumcentre do coincide, i. e., the triangle is equilateral.

1989/3. *Let n and k be positive integers and let S be a set of n points in the plane such that no three points of S are collinear, and for any point P of S there are at least k points of S equidistant from P. Prove that*

$$k < \frac{1}{2} + \sqrt{2n}.$$

First solution. Rearranging and squaring the given inequality we get $\left(k - \frac{1}{2}\right)^2 = k^2 - k + \frac{1}{4} < 2n$ which, when divided by 2, becomes

$$\frac{k^2 - k}{2} + \frac{1}{8} = \binom{k}{2} + \frac{1}{8} < n.$$

Hence it is enough to show that $\binom{k}{2} \le n - 1$.

For each point P of the set S consider k points, also in S and being equidistant from P, by condition. There are $\binom{k}{2}$ pairs formed from these points and hence there are

(1) $$n \cdot \binom{k}{2}$$

pairs listed, altogether.

If a certain pair (A, B) is checked at point P then, of course, P is on the perpendicular bisector of AB. Since there are no three collinear points in S, pair (A, B) has been counted at most twice in (1). Product (1), hence, cannot exceed the double of the total number of pairs formed by the points of S:

(2) $$n \cdot \binom{k}{2} \le 2\binom{n}{2}.$$

The *r.h.s.* is $n(n-1)$ and dividing by n yields the claim.

Second solution. Surprisingly enough the first condition about collinearity can be ignored. Denote the set of pairs formed by those points in S which are equidistant from $P_i \in S$ by H_i. If its cardinality is the usual $|H_i|$ then the condition is

(3)
$$|H_i| \geq \binom{k}{2}.$$

The points of the pairs in H_i are on a circle of centre P_i and those of H_j are on another one about P_j. These two circles have at most two common points that makes a single pair. Thus

(4)
$$|H_i \cap H_j| \leq 1.$$

Notice now that the cardinality of the union of the sets H_i when taken for every point $P_i \in S$ can be estimated from below as

(5)
$$\left| \bigcup_{i=1}^{n} H_i \right| \geq \sum_{i=1}^{n} |H_i| - \sum_{i,j} |H_i \cap H_j|.$$

This is a simple corollary of the *Principle of Inclusion and Exclusion* but it is straightforward anyway: the pairs occuring in more H_i at the *r.h.s.* have been subtracted more than once.

By (3) the terms of the first sum are at least $\binom{k}{2}$. In the subtracted sum in (5) there are $\binom{n}{2}$ terms, at most 1 each. Hence

$$\left| \bigcup_{i=1}^{n} H_i \right| \geq n\binom{k}{2} - \binom{n}{2}.$$

On the *l.h.s.* of (5), on the other hand, certain pairs, formed by the n points of S, have been counted, their total is hence at most $\binom{n}{2}$. Putting the estimates together:

$$\binom{n}{2} \geq n\binom{k}{2} - \binom{n}{2}, \quad \text{that is} \quad 2\binom{n}{2} \geq n\binom{k}{2},$$

which completes the proof.

1989/4. *Let $ABCD$ a convex quadrilateral such that the sides AB, BC, AD satisfy $AB = AD + BC$. There exists a point P inside the quadrilateral at a distance h from the line CD such that $AP = h + AD$ and $BP = h + BC$. Show* *that*

(1)
$$\frac{1}{\sqrt{h}} \geq \frac{1}{\sqrt{AD}} + \frac{1}{\sqrt{BC}}.$$

Solution. Draw circles k_1 and k_2 of radii $AD = r_1$ and $BC = r_2$ about the vertices A and D, respectively. According to the conditions these circles are touching each other externally inside AB and circle k of radius h about P is touching both of them and also the side CD (*Figure 1989/4.1*).

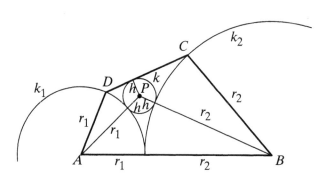

Figure 89/4.1

Consider now the varying quadrilaterals while the vertices A, B and also the two circles k_1 and k_2 are kept fixed ($r_2 \geq r_1$ can clearly be assumed). In these quadrilaterals CD is connecting two points of the respective circles and k, the varying circle is touching this segment. It is obvious that k can increase until it is touching the common exterior tangent to the circles k_1 and k_2 and this limiting position yields the highest value of h. Denote this value by h_1 clearly $h_1 \geq h$ for any possible h, that is $\sqrt{h_1} \geq \sqrt{h}$ or

(2)
$$\frac{1}{\sqrt{h}} \geq \frac{1}{\sqrt{h_1}}.$$

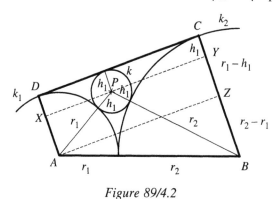

Figure 89/4.2

Hence it is enough to prove (1) for this extremal value h_1 when C and D are on the common tangent of k_1 and k_2. Quadrilateral $ABCD$ is now a right angled trapezoid. Denote the projections of P on AD and BC by X and Y respectively and that of A on BC by Z. By Pythagoras in the right triangles PXA

and PYB (*Figure 1989/4.2*)

$$XP = \sqrt{(r_1 + h_1)^2 - (r_1 - h_1)^2} = 2\sqrt{r_1 h_1},$$

$$PY = \sqrt{(r_2 + h_1)^2 - (r_2 - h_1)^2} = 2\sqrt{r_2 h_1}$$

and also in the right triangle AZB

$$AZ = \sqrt{(r_2 + r_1)^2 - (r_2 - r_1)^2} = 2\sqrt{r_1 r_2}.$$

Since $AZ = XP + PY$,

$$2\sqrt{r_1 r_2} = 2\sqrt{r_1 h_1} + 2\sqrt{r_2 h_1}.$$

Dividing by $2\sqrt{r_1 r_2 h_1}$

$$\frac{1}{\sqrt{h_1}} = \frac{1}{\sqrt{r_1}} + \frac{1}{\sqrt{r_2}},$$

which, by(2), implies

$$\frac{1}{\sqrt{h}} \geq \frac{1}{\sqrt{AD}} + \frac{1}{\sqrt{BC}},$$

indeed, and it also follows that equality holds if and only if $ABCD$ is a right angled trapezoid.

1989/5. *Prove that for each positive integer n there exist n consecutive positive integers none of which is a prime or a prime power.*

First solution. For a given n we are looking for that value of M for which none of the n consecutive numbers

(1) $$2 + M, \; 3 + M, \; \ldots, \; (n+1) + M$$

is a prime power. An integer k is not a prime or the power of a prime if it has two coprime divisors both less than k.

Let $i + M$ be a term in (1). If i divides M then $m_i M = i m_i$ for some positive integer or

$$i + M = i(1 + m_i).$$

The crucial observation is that if, additionally, i also divides m_i that is M is a multiple of i^2 then i and $1 + m_i$ have no common factor: i and $1 + i m_i$ are coprime factors of $M + i$, as required.

A number of the given property is clearly $M = ((n+1)!)^2$ yielding the n element list

$$2 + ((n+1)!)^2, \; 3 + ((n+1)!)^2, \; \ldots, \; (n+1) + ((n+1)!)^2,$$

with no prime powers, indeed.

Second solution. The claim clearly holds for $n = 1$ and 2; for $n > 2$ let p_1, p_2, $\ldots$, p_k primes not exceeding n (that is $p_1 = 2$, $p_2 = 3$, $p_3 = 5$, $\ldots$) and with

$N = p_1 p_2 \ldots p_k$ consider the list

(2) $\qquad\qquad N!+1, \; N!+2, \; \ldots, \; N!+n$

of n consecutive integers. We prove that none of these numbers is a prime power. Assume the contrary, that is some $N!+j$, $1 \le j \le n$ is equal to the rth power of a prime p,

(3) $\qquad\qquad\qquad N!+j = p^r.$

Since $N > n \ge j$, this j is among the factors $N!$ and thus the *l.h.s.* of (3) implies that j is a divisor of p^r. This j, hence, is equal to p^s, where $s < r$ is a positive integer. simplifying by j in (3) yields

(4) $\qquad 1 \cdot 2 \cdot 3 \cdot \ldots \cdot (j-1) \cdot (j+1) \cdot \ldots \cdot N + 1 = p^{r-s}.$

$j = p^s$, on the other hand, implies that $p \le j \le n$, p, as a prime not exceeding n is one of those listed as $p_1, p_2, \ldots, p_k$. By construction, however, p then divides N, a contradiction.

Note. In the solution we used the estimate $N > n$. This should, in fact, be proved. Assuming the contrary, $N = p_1 p_2 \ldots p_n \le n$, none of the primes $p_1, p_2, \ldots, p_k$ can divide the one less $N - 1$. Since these are all the primes up to n, $N - 1$ as a number less than n cannot have prime divisors at all, it must be equal to 1. This means that $k = 1$ and $n = 2$ which was excluded at the beginning.

1989/6. *A permutation* $x_1, x_2, \ldots, x_{2n-1}, x_{2n}$ *of the set* 1, 2, $\ldots$, $2n - 1$, $2n$ *where n is a positive integer is said to have property P if* $|x_i - x_{i+1}| = n$ *for at least one i in* $\{1, 2, \ldots, 2n-1\}$. *Show that for each n there are more permutations with property P than without.*

First solution. Call, for brevity, permutations with property P simply P-permutations and the remaining ones Q-permutations. For any $1 \le i \le 2n$ set j as the pair of i if $1 \le j \le 2n$ and $|i - j| = n$, that is the numbers whose difference is n are arranged into pairs.

We shall prove the claim by defining a simple one to one mapping from the set of Q-permutations to a proper subset of the P-permutations. If $x_1, x_2, \ldots, x_{2n}$ is a Q-permutation and the pair of x_{2n} is x_k then define the image of this permutation to be the one whose last element has been removed and inserted behind its pair. This image is clearly a P-permutation and, additionally, it contains but a single neighbouring pair that is not at the end of the permutation.

It is clear that from any P-permutation of this property we can reconstruct the corresponding Q-permutation, the mapping is indeed one to one. Going on, if $n > 1$ then 1, 2, 3, $\ldots$, $n-1$, $n+1$, $\ldots$, n, $2n$ is a P-permutation and not like those in the range of our mapping; it is hence a proper subset of the P-permutations, indeed. Finally, if $n = 1$ then there is just one permutation and it has property P, the proof is hence complete.

Second solution. Denote the number of those permutations of the $2n$ numbers that contain i and $i+n$ next to each other by $\mathcal{P}_i$. These sets consist of P-permutations only, their union is clearly the set of P-permutations that is denoted by $\mathcal{P}$. With usual notations

(1)
$$|\mathcal{P}| = \left|\bigcup_{i=1}^{n} \mathcal{P}_i\right| \geq \sum_{i=1}^{n} |\mathcal{P}_i| - \sum_{i<k} |\mathcal{P}_i \cap \mathcal{P}_k|.$$

This is again a corollary of the *Principle of Inclusion and Exclusion* already used in the Second solution of problem 3.

Now $|\mathcal{P}_i| = 2 \cdot (2n-1)!$ because the elements of $\mathcal{P}_i$ can clearly be obtained by treating the pairs $(i, i+n)$ and $(i+n, i)$ as single elements and preparing the $(2n-1)!$ permutations of the hence obtained $2n-1$ elements.

Proceed likewise when counting the intersection $\mathcal{P}_i \cap \mathcal{P}_k$: 'glue' the numbers i and $i+n$ and also k and $k+n$. These pairs form $(2n-2)!$ permutations with the remaining $2n-4$ single numbers. Since there are 4 possible orders of the numbers in the two pairs

$$|\mathcal{P}_i \cap \mathcal{P}_k| = 4 \cdot (2n-2)! \ .$$

Remembering that there are $\binom{n}{2}$ pairwise formed intersections, (1) becomes

$$|\mathcal{P}| \geq n \cdot 2 \cdot (2n-1)! - \binom{n}{2} \cdot 4 \cdot (2n-2)! =$$
$$= (2n)! - \frac{2n(2n-1)(2n-2)!(2n-2)}{2(2n-1)} = (2n)! - \frac{(2n)!(2n-2)}{2(2n-1)} =$$
$$= (2n)! \left(1 - \frac{1}{2} \cdot \frac{2n-2}{2n-1}\right) > \frac{(2n)!}{2}.$$

This, in fact, is nothing else but the assertion of the problem.

Note. Estimate (1) in the Second solution can be improved and then we get that the number of P-permutations when divided by the total number of permutations tends to the ratio

$$\frac{e-1}{e} = 0.63212055\ldots$$

as n tends to infinity; for sufficiently large values of n, hence, roughly 63.21 % of the permutations have property P.

1990.

1990/1. *Chords AB and CD of a circle intersect at a point E inside the circle. Let M be an interior point of the segment EB. The tangents to the circle through D, E and M intersects the lines BC and AC at F and G respectively. Find $\dfrac{EG}{EF}$ in terms of $t = \dfrac{AM}{AB}$.*

Solution. A careful check of the diagram (*Figure 1990/1.1*) reveals quite a number of equal angles. By the theorem of intercepted angles

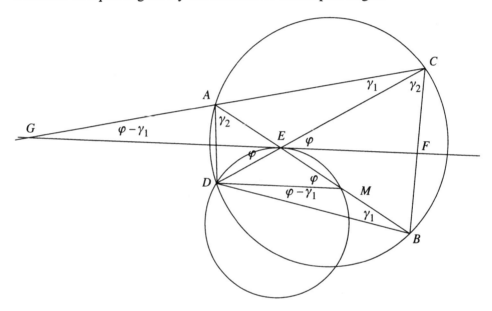

Figure 90/1.1

$$\angle ACD = \angle ABD = \gamma_1, \qquad \angle DAB = \angle DCB = \gamma_2,$$
$$\angle DME = \angle DEG = \angle FEC = \varphi.$$

Simple calculation shows that $\angle CGF = \angle MDB = \varphi - \gamma_1$.

The equality of the respective angles now implies that $\triangle CEF$ and $\triangle AMD$ and triangles CGE, BDM are pairwise similar. Being so

(1) $\qquad \dfrac{CE}{EF} = \dfrac{AM}{DM}, \qquad$ that is $\qquad CE \cdot DM = AM \cdot EF,$

and also

(2) $\qquad \dfrac{CE}{GE} = \dfrac{MB}{DM}, \qquad$ that is $\qquad CE \cdot DM = GE \cdot MB.$

Combining (1) and (2) yields $AM \cdot EF = GE \cdot MB$, or $AM = t \cdot AB$. Using $MB = AB - AM = (1 - t)AB$ we arrive to the desired result:

$$\frac{GE}{EF} = \frac{AM}{MB} = \frac{t \cdot AB}{(1 - t)AB} = \frac{t}{1 - t}.$$

Remark. In the argument we used that A separates G and C. This order of the points clearly follows from $\varphi > \gamma_1$.

1990/2. *Take $n \geq 3$ and consider a set E of $2n-1$ distinct points on a circle. Suppose that exactly k of these points are to be coloured black. Such a colouring is "good" if there is at least one pair of black points such that the interior of one of the arcs between them contains exactly n points from E. Find the smallest value of k so that every such colouring of k points of E is good.*

Solution. We may assume that our points, $A_1, A_2, \ldots, A_{2n-1}$ are the vertices of a regular $(2n-1)$-gon. In what follows, distances between these vertices will be measured by one of the connecting arcs specified in due course. Setting the circular distance between two adjacent vertices as the unit, some of these distances might turn out to be greater than $2n-1$, that is the total perimeter; the actual distance then is clearly the residue when dividing by $2n-1$; subscripts should be reduced likewise, if necessary.

An arc between two vertices contains n points if its length is $n+1$; there are two points this far from A_{n+1}, for example: $A_{2(n+1)} = A_3$ and A_{2n-1}. Start now from A_{n+1} and mark, in one direction, the points being $(n+1)$ far from the previous one:

(1) $A_{n+1}, \; A_{2(n+1)}, \; A_{3(n+1)}, \; \ldots, \; A_{c(n+1)},$

where c is the highest integer for which the points above are all distinct, the length of the cycle in (1). Thus the next point, $A_{(c+1)(n+1)}$ appears on the list already, what's more, it is the very starting point, A_{n+1}. Indeed, the list contains both $(n+1)$-neighbours of any other point. Being so

$$(c+1)(n+1) - (n+1) = t(2n-1),$$

for some positive integer t. Rearranging we get

(2) $c(n+1) = t(2n-1).$

Consider first when the cycle is maximal that is $c = 2n-1$. With $t = n+ +1$ they satisfy (2) and now (1) includes every point of E. If n of them are coloured then (1) contains at least one adjacent black pair. Since the length of their arc along the chosen direction is $n+1$ there are n points from E on this arc, indeed. With fewer points painted, on the other hand, it is clearly possible to avoid neighbouring black ones in (1). Thus if the cycle contains every point then the smallest value of k is n.

The cycle certainly locks at the $(2n-1)$th step but this can already occur for smaller values of c. Let d be the g.c.d. of $n+1$ and $2n-1$. Then d divides their difference, $2(n+1) - (2n-1) = 3$ that is either $d=1$ or $d=3$. If $d=3$ then the subscript of every vertex in (1) is a multiple of 3 and thus the list itself does not exhaust the set E; if the cycle is maximal then d is forced to be 1. If $d=3$

then $2n - 1$ as an odd multiple of 3 is equal to $3(2s - 1) = 6s - 3$, and hence $n + 1 = 3s$. Substituting into (2):

$$3sc = t(6s - 3), \quad \text{or} \quad sc = t(2s - 1).$$

$c = 2s - 1 = \dfrac{2n - 1}{3}$ satisfies this equation and thus the following $2s - 1$ points

(3) $$A_{n+1}, \; A_{2(n+1)}, \; \ldots, \; A_{\frac{2n-1}{3}(n+1)}$$

are all distinct. Repeating the previous argument (then s was equal to n) we get that $s - 1$ can still be painted black without having adjacent black points in (3) but this cannot be done any more if there are s points to be coloured.

Rotating the set (3) by one and two units respectively we obtain the points of E with subscripts $3r + 1$ and $3r + 2$ and these points, together with (3) exhaust the set E. In both sequences as in (3) one can colour $s - 1$ but no more points without having adjacent pairs. Hence the maximal number of black points where there are no two points of the given property is $3(s - 1) = n - 2$, that is no matter how we colour $n - 1$ points black there will always be two points as required.

Coming to the end: if the g.c.d. of $2n - 1$ and $n + 1$ is equal to 1 then $k = n$; if it is equal to 3 then $k = n - 1$.

This can also be put as follows:

if 3 does not divide $2n - 1$ then $k = n$;

if 3 divides $2n - 1$ then $k = n - 1$.

1990/3. *Determine all integers greater than 1 such that* $\dfrac{2^n + 1}{n^2}$ *is an integer.*

Solution. $n = 3$ is a trivial solution as $\dfrac{2^3 + 1}{3^2} = 1$. We show that there is no other solution. We shall proceed by proving a series of lemmas.

Lemma 1. If k is odd then $s = 2^{2k} - 2^k + 1$ is divisible by 3 but not by 9.

For odd powers of 2

$$2^1 \equiv 2, \quad 2^3 \equiv 8, \quad 2^5 \equiv 5, \quad 2^7 \equiv 2, \quad 2^9 \equiv 8, \quad \ldots \quad (\text{mod } 9)$$

and the remainder, in general, of 2^{6k+r} ($r = 1, 3, 5$) when divided by 9 depends on r only, namely

$$2^{6k+r} = (64)^k \cdot 2^r \equiv 2^r \quad (\text{mod } 9).$$

Hence

$$2^{6k+1} \equiv 2, \quad (2^{6k+1})^2 \equiv 4, \quad s \equiv 4 - 2 + 1 = 3,$$

$$2^{6k+3} \equiv 8, \quad (2^{6k+3})^2 \equiv 1, \quad s \equiv 1 - 8 + 1 = 3, \quad (\text{mod } 9)$$

$$2^{6k+5} \equiv 5, \quad (2^{6k+5})^2 \equiv 7, \quad s \equiv 7 - 5 + 1 = 3.$$

Thus, for every odd value of k, the corresponding power of 2 is of the form $3 + 9i$, a multiple of 3 but not that of 9, indeed.

Lemma 2. If a, b and p are positive integers such that $2^a \equiv 2^b \equiv 1$ (mod p) and $m = (a, b)$ then $2^m \equiv 1$ (mod p).

Let $a = bq + r$ $(0 \le r < b)$. Then

(1)
$$1 \equiv 2^a = 2^{bq+r} = \left(2^b\right)^q \cdot 2^r \equiv 2^r \quad (\text{mod } p)$$

Thus if m is computed by *Euclid's algorithm* then congruence (1) holds for the stepwise computed remainders and, in particular, for the last one different from zero, the greatest common divisor of a and b:

(2)
$$2^m \equiv 1 \quad (\text{mod } p).$$

Lemma 3. If n is a positive integer then $2^n + 1$ is not divisible by 7.

This is true if $n = 1$, 2, 3. Assume that the lemma is false and there exists some $n > 3$ for which $2^n + 1$ is a multiple of 7. Denote the smallest index of this property by n_0. Then, for some positive integer N we have $7N = 2^{n_0} + 1$. But now

$$7(N+1) = 2^3 \cdot 2^{n_0-3} + 1 + 7 = 8\left(2^{n_0-3} + 1\right),$$

$2^{n_0-3} + 1$ is also divisible by 7 contradicting to the choice of n_0. The proof of the lemma is hence complete.

Let's turn now to the actual solution. $2^n + 1$ is odd and so is its divisor, n. Let's write it as $n = 3^k \cdot d$ where $k \ge 0$ integer and 3 does not divide the odd d. If $\dfrac{2^n + 1}{n^2}$ is an integer then using the identity $a^3 + 1 = (a+1)(a^2 - a + 1)$ it follows, by induction, that

$$a^{3^k} + 1 = (a+1) \prod_{i=0}^{k-1} \left(a^{2 \cdot 3^i} - a^{3^i} + 1\right).$$

Substituting 2^d for a and using that $n = 3^k d$ we get

(1)
$$2^n + 1 = \left(2^d + 1\right) \prod_{i=0}^{k-1} \left(2^{2 \cdot 3^i d} - 2^{3^i d} + 1\right).$$

By Lemma 1. the 3-term factors on the *r.h.s.* are divisible by exactly the first power of 3 and hence the index of 3 in the product is k. This product, by condition, is divisible by n^2 and thus by 3^{2k}; $2^d + 1$, hence, is divisible by 3^k. The proof of Lemma 1. also implies that the index of 3 in $2^d + 1$ is at most 1. Thus either $k = 0$ or $k = 1$ that is

(2)
$$n = d \quad \text{or} \quad n = 3d.$$

It is enough to show that $d = 1$. Assume the contrary, $d > 1$, and denote the smallest prime divisor of d by p. By now we know that p is different from both 2 and 3 so it is at least 5. By its choice $p - 1$ and d have no common factor,

$(p-1,d)=1$. We also know that p is a factor of n and hence it also divides 2^n+1 that is

(3) $$2^n \equiv -1 \pmod{p}.$$

Squaring

$$2^{2n} \equiv 1 \pmod{p},$$

on the other hand, by *Fermat's theorem*

$$2^{p-1} \equiv 1 \pmod{p}.$$

Applying Lemma 2. to the latter two congruences yields

(4) $$2^m \equiv 1 \pmod{p},$$

with $m=(2n,p-1)$ and thus, by (2) m divides $6d$. Since m divides $p-1$ which has no common factor with d we have succeeded to reduce the possible values of m: it must divide 6. Hence, by (4), p is a prime factor, greater than 3, of one of the numbers 1, 3, 7, 63. The only possibility is $p=7$ and this, by (3), implies that 7 is a proper divisor of 2^n+1. This, however, contradicts to Lemma 3 and thus d, in (2), is indeed equal to 1. Since $n>1$, the only solution is $n=3$, the proof is finished.

1990/4. *Construct a function from the set of positive rational numbers into itself such that*

(1) $$f(x \cdot f(y)) = \frac{f(x)}{y}$$

for all x, y.

Solution. First we note that if a function f satisfies the functional equations

(2) $f(f(x)) = \dfrac{1}{x}$ and (3) $f(x)f(y) = f(xy)$

then it also satisfies (1) and thus it is enough to find a functional solution of equations (2) and (3). The solutions of (3), by the way, are the so called *multiplicative functions* and the property given there is also called *multiplicativity*.

Denote, for the construction, the ith prime by p_i: $p_1=2$, $p_2=3$, ... Any rational number can be written as

$$x = p_1^{\alpha_1} p_2^{\alpha_2} \cdots p_{2n-1}^{\alpha_{2n-1}} p_{2n}^{\alpha_{2n}} \cdots$$

where α_i is positive or negative integer, maybe zero. The index of p_i is zero after a certain position anyway and the corresponding primes can be omitted from the product above. Set now

(4) $$f(x) = p_1^{\alpha_2} p_2^{-\alpha_1} \cdots p_{2n-1}^{\alpha_{2n}} p_{2n}^{-\alpha_{2n-1}} \cdots .$$

This mapping works by grouping the prime numbers into pairs, swapping the indices within each pair, finally, the index thus obtained is multiplied by -1 if it is at even position. By the unique prime factorisation it is enough to verify multiplicativity if the arguments are prime numbers and this is obvious for the 'prime-wise' acting function defined in (4). To prove that f satisfies (2) observe that

$$f(f(x)) = p_1^{-\alpha_1} p_2^{-\alpha_2} \cdots p_{2n-1}^{-\alpha_{2n-1}} p_{2n}^{-\alpha_{2n}} = \frac{1}{x},$$

which is but equation (2). Satisfying both (2) and (3) function (4) indeed has the desired property.

Note. Those who have solved the problem usually realized that (1) also implies (2) and (3). This is not necessary for the solution, however, we still present a proof. Setting $x = y = 1$ in (1) yields

$$f(f(1)) = f(1), \quad \text{that is} \quad f(f(f(1))) = f(f(1)) = f(1).$$

Substitute now 1 for x and $f(1)$ for y:

$$f(f(f(1))) = \frac{f(1)}{f(1)} = 1.$$

Combining the results we obtain that $f(1) = 1$.

Using (1) again:

$$f(f(y)) = f(1 \cdot f(y)) = \frac{f(1)}{y} = \frac{1}{y},$$

that is

(5)
$$f(f(y)) = \frac{1}{y},$$

which is (2), indeed. f when acting on both sides of (5) yields the following equality:

(6)
$$f(f(f(y))) = f\left(\frac{1}{y}\right).$$

Substituting $f(y)$ for y in (5) and comparing the result with (6) we get

$$f(f(f(y))) = \frac{1}{f(y)},$$

$$f\left(\frac{1}{y}\right) = \frac{1}{f(y)}.$$

Replacing y by $f(y)$ again and applying (5):

$$f\left(\frac{1}{f(y)}\right) = \frac{1}{f(f(y))} = y.$$

Finally, evaluating (1) once more and using the last equality we get

$$f(xy) = f\left(xf\left(\frac{1}{f(y)}\right)\right) = \frac{f(x)}{\frac{1}{f(y)}} = f(x)f(y);$$

we have arrived at (2) and this is the end.

1990/5. *Given an initial integer $n_0 > 1$, two players A and B choose integers n_1, n_2, n_3, ... alternately according to the following rules: knowing n_{2k}, A chooses any integer n_{2k+1} such that*

$$n_{2k} \leq n_{2k+1} \leq n_{2k}^2.$$

Knowing n_{2k+1}, B chooses any integer n_{2k+2} such that

$$\frac{n_{2k+1}}{n_{2k+2}} = p^r$$

for some prime p and integer $1 \leq r$. Player A wins the game by choosing the number 1990; player B wins by choosing number 1. For which n_0 does

a) *A have a winning strategy?*

b) *B have a winning strategy?*

c) *Neither player have a winning strategy?*

Solution. Suppose that the starting value n_0 has just been announced and it is now player A's turn.

If $45 \leq n_0 \leq 1990$ then $1990 < 45^2 = 2025$ and thus A can choose $n_1 = 1990$ and win immediately.

If $27 \leq n_0 \leq 44$ then $n_1 = 720 = 2^4 \cdot 5$ is a legal choice for A. To make sure that $\frac{n_1}{n_2}$ is a prime power B has to choose n_2 between 45 and 360 and thus A can pick 1990 as n_3 because $n_2 \leq 1990 \leq n_2^2$.

If $15 \leq n_0 \leq 26$ then $n_1 = 210 = 2 \cdot 3 \cdot 5 \cdot 7$ is available for A; player B is now allowed to reply in the range $[30, 105]$ only and hence A can finish by choosing 1990 again.

If $11 \leq n_0 \leq 14$ then A should choose $n_1 = 105 = 3 \cdot 5 \cdot 7$. The range for B is now $15 \leq n_2 \leq 35$ but these values have already been verified to be in favour of A.

If $8 \leq n_0 \leq 10$ then, similarly, if A starts with $n_1 = 60 = 2^2 \cdot 3 \cdot 5$ then $12 \leq n_2 \leq \leq 30$ so, like before, A is to win.

If $n_0 < 8$ then the triumphants of A are over: none of his options would force B beyond 7.

The state of affairs so far is that A has a winning strategy as long as $8 \leq n_0 \leq \leq 1990$.

If $2 \leq n_0 \leq 5$ then player B can win and there is no better way to prove this but checking the starting values one by one.

If $n_0 = 2$ then n_1 can be only 2, 3 or 4 and thus B can respond by saying 1 and hence win.

If $n_0 = 3$ then the legal values of n_1 are 3, 4, 5, 6, 7, 8, 9; if A chooses $n_1 = 6$ then B can reply with $n_2 = 2$-t and can win as before; for any other choice of A player B can prompt 1.

If $n_0 = 4$ then if n_1 is from the list $\{4, 5, 6, 7, 8, 9\}$ then the players are back in the previous case; the values remaining are 10, 11, 12, 13, 14, 15, 16 and the winning replies of B are $n_2 = 2, 1, 3, 1, 2, 3, 1$ respectively.

If $n_0 = 5$ then the range $n_1 \leq 16$ has already been checked. For the remaining values

B can choose
$$n_1 = 17, \quad 18, \quad 19, \quad 20, \quad 21, \quad 22, \quad 23, \quad 24, \quad 25,$$
$$n_2 = 1, \quad 2, \quad 1, \quad 4, \quad 3, \quad 2, \quad 1, \quad 3, \quad 1$$

respectively and thus each case is reduced to some previously checked position all in favour of B.

There are two cases left, $n_0 = 6$ and $n_0 = 7$.

If $n_0 = 6$ then A can reply from the range $6 \leq n_1 \leq 36$. Apart from $n_1 = 30$ B can always choose one of the numbers 1, 2, ..., 5 as n_2 and, as we have already seen, win afterwards. To avoid defeat A has only one choice, namely $n_1 = 30$. Now it is B in a similar dead end street: among the divisors of 30 he should choose one which is less than 8 but there is only one such number left: $n_2 = 6$. Mutually forced to play against defeat they are back at beginning; from now on the game goes on and on, indefinitely, none of the players has a winning strategy.

The story is similar if $n_0 = 7$. The numbers for A are now $7 \leq n_1 \leq 49$. If n_1 is different from both 30 and 42 then B can always choose some of his winning numbers 1, 2, ..., 5. $n_1 = 30$ forces the infinite game above but the game is also blocked if A chooses $n_1 = 42$ because the only reasonable answer of B is the vicious $n_2 = 6$. Coming to the end: if $n_0 = 6$ or 7 then none of the players can win the game.

We show that A can also win if $n_0 > 1990$. Consider the interval $(n_0, 2n_0)$ for $n_0 > 1990$. It contains at least 1991 whole numbers and hence there are certainly some of them of the form $9 \cdot 2^k$, $(k > 3)$. Let A choose one of these numbers i.e., let $n_1 = 9 \cdot 2^k$. Now B's reply, n_1, is at least 9 but not greater than $9 \cdot 2^{k-1}$; B is forced to choose his number below n_0. Following this strategy A can push his opponent below 1990 in finitely many steps keeping him above 9 all the time. Since this range was proved to be in favour of A he can indeed win the game if n_0 is greater than 1990.

Summing up:

A has a winning strategy if $n_0 \geq 8$;

B has a winning strategy if $2 \leq n_0 \leq 5$;

neither player has a winning strategy if $n_0 = 6$ or 7.

Remark. The winning strategy of A is not unique; playing differently he can also win the game. This, of course, has no effect on the conclusion.

1990/6. *Prove that there exists a convex 1990-gon such that all its angles are equal and the lengths of the sides are the numbers 1^2, 2^2, ..., 1989^2, 1990^2 in some order.*

Solution. The sides of the convex polygon to be constructed are parallel to those of a regular 1990-gon respectively. First we construct a $995 = \dfrac{1990}{2}$ sided polygon of suitable sides then each side will be divided into two parts and these parts when rotated in proper order will form the required polygon.

Working on the complex plane will make our life easier. Let e be the first 995th root of unity:

$$e = \cos \frac{2\pi}{995} + i \sin \frac{2\pi}{995}.$$

This number obviously satisfies $e^{995} = 1$ that is $e^{995} - 1 = 0$. Hence

$$(e - 1)\left(1 + e + e^2 + \ldots + e^{993} + e^{994}\right) = 0,$$

(1) $$\qquad S_1 = 1 + e + e^2 + \ldots + e^{993} + e^{994} = 0,$$

on the other hand $\left(e^5\right)^{199} - 1 = 0$ implies

$$\left(e^5 - 1\right)\left(1 + e^5 + e^{10} + \ldots + e^{985} + e^{990}\right) = 0,$$

(2) $$\qquad S_2 = 1 + e^5 + e^{10} + \ldots + e^{985} + e^{990} = 0.$$

The sides of a regular 995-gon are parallel to the terms of (1) respectively. Prepare now the following linear combination of these numbers/vectors:

(3) $$S = \sum_{b=0}^{198} \Big(b \cdot e^{5b} + (b+199)e^{5b+199} + (b + 2 \cdot 199)e^{5b+2 \cdot 199} +$$

$$+ (b + 3 \cdot 199)e^{5b+3 \cdot 199} + (b + 4 \cdot 199)e^{5b+4 \cdot 199}\Big).$$

There are $5 \cdot 199 = 995$ terms in the sum S. The coefficients, $b + n \cdot 199$ ($n = 0$, 1, 2, 3, 4) are the integers from 0 to 994, each of them occurring once. We show that the complex terms are but the roots of unity in (1), remembering that $e^{995} = 1$. It is enough to show that the complex numbers — each of them is 995th root of unity — in (3) are all distinct. The opposite would imply either

$$5b + n \cdot 199 = 5b' + j \cdot 199, \quad \text{that is} \quad 5(b - b') = 199(j - n)$$

but then $b - b'$ would be a multiple of 199 and the latter can happen only if $b = b'$ and, accordingly, $j = n$; or

$$5b + n \cdot 199 = 5b' + j \cdot 199 + 995$$

which is likewise impossible. The roots of unity in (3) are hence distinct and as there are 995 of them they indeed exhaust the list in (1).

Next we prove that $S = 0$. Multiply for this by $1 - e^{199} \neq 0$. Performing the calculations we get

$$\left(1 - e^{199}\right) S = \sum_{b=0}^{198} e^{5b} \left(199 e^{199} \left(1 + e^{199} + e^{2 \cdot 199} + e^{3 \cdot 199} - 4 e^{4 \cdot 199}\right)\right).$$

Denote the 5-term sum in the brackets by E. Thus

$$\left(1 - e^{199}\right) S = E \sum_{b=0}^{198} e^{5b} = E \left(1 + e^5 + e^{10} + \ldots + e^{990}\right),$$

and, by (2), this is zero, indeed.

Rearrange now the terms in (3) in the natural order of the coefficients $b + 199 j = k$ $(j = 0, 1, 2, 3, 4)$ and denote the complex root of unity with coefficient k by e_k $(e_0 = 1)$. (3) then can be written as $S = \sum_{k=0}^{994} k e_k$. Remembering that the numbers e_k are the 995th roots of unity and thus their sum is zero, we now consider the sum

$$995 \cdot \left(997 \cdot \sum_{k=0}^{994} e_k + 2S\right) = 995 \cdot \left(997 \cdot \sum_{k=0}^{994} e_k + 2 \cdot \sum_{k=0}^{994} k e_k\right) = 0;$$

this, by the way, can also be written as

(4) $$\sum_{k=0}^{994} 995 \cdot (995 + 2(k+1)) e_k = 0.$$

Since $995(995 + 2(k+1)) = (995 + k + 1)^2 - (k+1)^2$, (4) is the same as

(5) $$\sum_{k=0}^{994} \left((995 + k + 1)^2 e_k + (k+1)^2 (-e_k)\right) = 0.$$

The complex numbers e_k together with their respective opposites $-e_k$ make up the 1990th roots of unity. Formula (5) shows that the terms when multiplied by $1^2, 2^2, \ldots, 1990^2$ respectively and in the order as they appear in the sum, form, in fact, a closed polygon whose sides are of the required length and direction. Finally if the order of the terms is matched to the ascending order of their respective arguments then the polygon obtained will be convex and thus the construction is complete.

7. A Glossary of Theorems

[1] *The paralelogram theorem and an application.* The sum of the squares of a paralelogram's diagonals is equal to that of the sides. Denote, for the proof, the vectors spanning the paralelogram by **a** and **b**; hence its diagonals are $\mathbf{a}+\mathbf{b}$ and $\mathbf{a}-\mathbf{b}$, respectively. The claim is now straightforward as

$$2\mathbf{a}^2+2\mathbf{b}^2=(\mathbf{a}+\mathbf{b})^2+(\mathbf{a}-\mathbf{b})^2.$$

Let the lengths of the sides of a triangle be a, b and c and that of the median $CC'=s_c$ (*Figure 1.1.*). Reflecting the triangle in C' yields a paralelogram of sides a, b, a, b; its diagonals are c and $2s_c$. By the previous result

$$2a^2+2b^2=4s_c^2+c^2,$$

and hence

$$s_c^2=\frac{2a^2+2b^2-c^2}{4}.$$

[2] *Sides and cotangents in a triangle.*

$$a^2+b^2+c^2=4t(\cot\alpha+\cot+\cot\gamma).$$

Write down the cosine rule for the sides and sum the equalities:

$$a^2+b^2+c^2=2\left(a^2+b^2+c^2\right)-2bc\cos\alpha-2ca\cos\beta-2ab\cos\gamma,$$

$$a^2+b^2+c^2=2\left(bc\cos\alpha+ca\cos\beta+ab\cos\gamma\right).$$

From the area formula: $bc=\dfrac{2A}{\sin\alpha}$; substituting the symmetric permutations of this relation into the previous result yields the claim. Indeed

$$a^2+b^2+c^2=4t\left(\frac{\cos\alpha}{\sin\alpha}+\frac{\cos\beta}{\sin\beta}+\frac{\cos\gamma}{\sin\gamma}\right)=4t(\cot\alpha+\cot\beta+\cot\gamma).$$

[3] *The cotangent inequality.* If α, β, γ are the angles of a triangle then

$$\cot\alpha+\cot\beta+\cot\gamma\geq\sqrt{3}.$$

There are several ways to prove this inequality; starting with straightforward identities we proceed by simple estimations.

$$\cot\alpha+\cot=\frac{\cos\alpha}{\sin\alpha}+\frac{\cos\beta}{\sin\beta}=\frac{\sin(\alpha+\beta)}{\sin\alpha\sin\beta}=\frac{2\sin\gamma}{\cos(\alpha-\beta)-\cos(\alpha+\beta)}=$$

$$=\frac{2\sin\gamma}{\cos(\alpha-\beta)+\cos\gamma}\geq\frac{2\sin\gamma}{1+\cos\gamma}=\frac{4\sin\frac{\gamma}{2}\cos\frac{\gamma}{2}}{1+2\cos^2\frac{\gamma}{2}-1}=2\tan\frac{\gamma}{2}.$$

Therefore

$$\cot\alpha+\cot\beta+\cot\gamma\geq\cot\gamma+2\tan\frac{\gamma}{2}=$$

$$=2\tan\frac{\gamma}{2}+\frac{\cot^2\frac{\gamma}{2}-1}{2\cot\frac{\gamma}{2}}=\frac{1}{2}\frac{\cot^2\frac{\gamma}{2}+3}{\cot\frac{\gamma}{2}}=\frac{1}{2}\left(\cot\frac{\gamma}{2}+3\tan\frac{\gamma}{2}\right).$$

169

The last sum can be estimated from below by the A.M.–G.M. inequality.

$$\cot\alpha + \cot\beta + \cot\gamma \geq \sqrt{\cot\frac{\gamma}{2} \cdot 3\tan\frac{\gamma}{2}} = \sqrt{3}.$$

[4] *Isogonal point* (isogonal = equiangular). The sides of a triangle are subtending equal angles at the isogonal point. This common angle is clearly equal to $120°$ and one can find such a point if the angles of the triangle are all less than $120°$.

The isogonal point is hence incident to the circular arcs through the endpoints of the sides corresponding to $120°$; these arcs do necessarily have a point in common.

The following simple construction is also leading to the isogonal point. Draw equilateral triangles ABC', BCA' and CAB' above the sides of the triangle ABC externally; the line segments AA', BB', CC' are then congruent (*Figure 4.1.*). The segments AA' and BB', for example, are equal because the rotation by $60°$ about C is mapping the triangle ACA' into $B'CB$; this is true if these triangles are degenerated into a segment.

If the isogonal point I does exist then it is incident to each of the segments AA', BB', CC'. In fact, since $AIC\angle = AIB\angle = 120°$ rotating the triangle AIB about A by $60°$ one obtains the triangle $AI'C'$. By the same rotation the triangle AII' is equilateral (*Figure 4.2.*). The diagram reveals that the points C, I and C' are collinear, therefore I is incident to the segment CC', indeed. Apart from that $CC' = IA + IB + IC$ and thus $CC'(= AA' = BB')$ is equal to the sum of the distances of the point I from the vertices of the triangle.

Starting with a point I not lying on the segment CC' the previous rotation yields the following inequality:

$$AI + BI + CI = II' + CI + I'C' > CC',$$

since the segments II', CI, $I'C'$ now form a triangle. Differently speaking, the isogonal point – if it exists – is minimizing the sum of the distances of a point from the vertices of a triangle; this minimum is actually equal to the common length of the segments AA', BB', CC'.

When written down in the triangles BCB', CAC', ABA', respectively, the cosine rule yields

$$BB' = CC' = AA' = a^2 + b^2 - 2ab\cos(\gamma + 60°) =$$
$$= b^2 + c^2 - 2bc\cos(\alpha + 60°) = c^2 + a^2 - 2ac\cos(\beta + 60°).$$

This latter equality, by the way, can be obtained in a straightforward manner without the above investigations.

[5] *Bicentric polygons; Poncelet's porisms.* A polygon is called bicentric if it has both an inscibed and a circumscribed circle. Any triangle is bicentric, for example, and also all the regular polygons. There is a relation, for bicentric polygons,

between the inradius r, the circumradius R and the distance d of the respective centres depending also on the number of sides. Here there is a list of the first few of them for 3, 4, 5 and 6 sided polygons.

$n = 3$: $d^2 = R^2 - 2Rr$ or: $\dfrac{1}{R+d} + \dfrac{1}{R-d} = \dfrac{1}{r}$,   (Euler)

$n = 4$: $\left(R^2 - d^2\right)^2 = 2r^2\left(R^2 + d^2\right)$ or: $\dfrac{1}{(R+d)^2} + \dfrac{1}{(R-d)^2} = \dfrac{1}{r^2}$,

$n = 5$: $r(R - d) = (R + d)iR - r - d\,(iR - r + d + i2R)$,

$n = 6$: $3\left(R^2 - d^2\right)^4 = 4r^2\left(R^2 + d^2\right)\left(R^2 - d^2\right) + 16R^2d^2r^4$.

The circles of bicentric polygons have the celebrated property that they are, in fact, shared by infinitely many bicentric $n$gons; to put it more precisely denote the incircle and the circumcircle of a bicentric polygon by c and C, respectively. Draw from an arbitrary point A_1 of C a tangent to c and let this line meet C at A_2. The – other – tangent from A_2 to c meets C at A_3, etc. The point A_{n+1} of this process is then incident to A_1, the sequence of segments hence drawn is closing up at the nth step yielding the porism of Poncelet in circles.

In the general form of Poncelet's theorem there are conic sections for circles; accordingly, the general theorem is a gem of projective geometry.

[6] *Power of a point with respect to a circle; radical axis; radical centre.* Given is the circle c of radius R and centre O, for an arbitrary point P in the plane of c the real number

$$h = PO^2 - R^2$$

is called the power of the point P with respect to the circle c. This number is positive, zero or negative if the point P is outside of c, lying on c or it is inside c. For external points the value of h is equal to the square of the tangent from P to the circle.

The locus of the points whose power is equal with repect to two non concentric circles is a straight line perpendicular to their axis. This line is called the radical axis of the two circles. It is the line of the common chord if the circles intersect; it is the common – internal – tangent if they touch; in general, it contains those points from where one can draw equal tangents to the circles.

In case of three circles the pairwise drawn radical axes are either parallel or they meet at a common point; in the latter case this point is called the radical centre of the three circles.

[7] *The position vector of the incentre.* Denote the position vectors of the vertices of the triangle ABC by **a**, **b** and **c**, respectively. Since the intersection C_1 of the bisector of the angle C and the opposite side AB is dividing AB in the ratio $b : a$

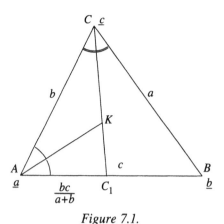

Figure 7.1.

(*Figure 7.1.*), the position vector of C_1 is equal to

$$\frac{a\mathbf{a}+b\mathbf{b}}{a+b}.$$

The bisector of the angle A meets the segment CC_1 at the incenter K dividing CC_1 in the ratio $b:\dfrac{bc}{a+b}=1:\dfrac{c}{a+b}$. K's position vector is hence

$$\mathbf{k}=\frac{\frac{c}{a+b}\cdot\mathbf{c}+\frac{a\mathbf{a}+b\mathbf{b}}{a+b}}{1+\frac{c}{a+b}}=\frac{a\mathbf{a}+b\mathbf{b}+c\mathbf{c}}{a+b+c}.$$

[8] *The inradius and the circumradius of a triangle.* There are several formulas for these radii in terms of the sides and the angles of the triangle; here there are but a few of them. The inradius is denoted by r, the circumradius is by R, the area of the triangle by A, the sides and the angles by a, b, c, and α, β, γ, respectively and finally, the semiperimeter by s.

$$R=\frac{abc}{4A}=\frac{a}{2\sin\alpha}=\frac{b}{2\sin\beta}=\frac{c}{2\sin\gamma},$$

$$R^2=\frac{2A}{\sin 2\alpha+\sin 2\beta+\sin 2\gamma}$$

$$r=\frac{A}{s}=4R\sin\frac{\alpha}{2}\sin\frac{\beta}{2}\sin\frac{\gamma}{2},$$

$$r^2=A\tan\frac{\alpha}{2}\tan\frac{\beta}{2}\tan\frac{\gamma}{2}.$$

[9] *Poncelet's theorem* see [5].

[10] *Touching circles.* Two circles in the space are touching each other if they have a single point in common and they share the tangent at this point. The axis of a circle is the line perpendicular to its plane at the centre. The axis contains the points whose distance from the points of the circle are all equal.

The plane perpendicular to the common tangent at the point T_{12} of contact of the touching circles c_1 and c_2 is containing the axes of both; if the circles are not coplanar then the axes meet and the distance of their common point O is equal to OT_{12} from the points of both circles; they are hence incident to the sphere of centre O and radius OT_{12}.

Consider now three pairwise touching circles c_1, c_2, c_3 that are not coplanar. Denote the touching points by T_{12}, T_{13}, T_{23}, respectively. Let the sphere containing c_1 and c_2 be S. Since there is a unique sphere passing through a circle c and a point not lying on c, the circle c_1 and the point T_{23} determine the sphere S; this sphere is also passing through the circle c_3. Accordingly, three pairwise touching circles, if not coplanar, are lying on a sphere.

11] *Equilateral tetrahedron* A tetrahedron is called equilateral if its faces are congruent. Being the face of an equilateral tetrahedron a triangle is always acute angled; any such triangle, on the other hand, is the face of some equilateral tetrahedron.

    The opposite edges of an equilateral tetrahedron are pairwise equal; its circumscribed parallelepiped is hence a cuboid. Its specific points, namely the incentre, the circumcentre and its centroid are incident; conversely, if any two of the above points are incident then the tetrahedron is equilateral. Finally, if the areas of the faces of a tetrahedron are equal then the terahedron is equilateral.

12] *The cosine inequality.* As a fundamental inequality in triangles it is the source of several further results; it states that

$$\cos \alpha + \cos \beta + \cos \gamma \le \frac{3}{2},$$

and equality holds if and only if the triangle is equilateral.

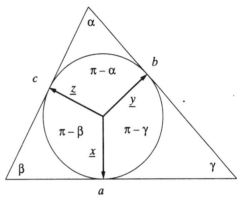

    The following non standard argument is using vectors: set the origin as the incentre and let the inradius be equal to 1. The unit vectors to the points of contact of the incircle are denoted by **x**, **y**, **z**, respectively. Using the notations of *Figure 12.1*

Figure 12.1.

$$0 \le (\mathbf{x}+\mathbf{y}+\mathbf{z})^2 = 3 + 2\mathbf{xy} + 2\mathbf{yz} + 2\mathbf{zx} =$$
$$= 3 + 2\,(\cos(a\pi - \gamma) + \cos(a\pi - \alpha) + \cos(a\pi - \beta)) = 3 - 2\,(\cos \alpha + \cos \beta + \cos \gamma),$$

yielding

$$\cos \alpha + \cos \beta + \cos \gamma \le \frac{3}{2},$$

and equality holds if and only if $\mathbf{x}+\mathbf{y}+\mathbf{z}=\mathbf{0}$, that is $1 = \mathbf{x}^2 = (-\mathbf{y}-\mathbf{z})^2 = 2 + {}$ $+ 2\cos(a\pi - \alpha)$, $\cos(a\pi - \alpha) = -\frac{1}{2}$, $180° - \alpha = 120°$. The pairwise angles of the vectors **x**, **y**, **z** are $120°$, the triangle is equilateral.

13] *Ramsey's theorem.* If the edges of an n-point graph are coloured with two colours say blue and red then Ramsey's theorem claims the existence of a monochromatic complete subgraph of certain size. Formally:

    For every pair (b, r) of positive integers there exists a positive integer $R(b, r)$ such that if $n \ge R(b, r)$ and each edge of a complete n-graph is coloured either blue or red then either there is a complete subgraph of b vertices whose edges are all blue or there is a complete subgraph of r vertices whose edges are all red. If, on the other hand, $n < R(b, r)$, then one can colour the edges of a complete

n-graph in such a way that there are no monochromatic complete subgraphs of the given sizes.

The task of finding the actual values of the so called Ramsey numbers $R(b,r)$ is extremely hard in general. There are some estimations but they are far from being sharp.

[14] *Position vectors of coplanar points.* Let the vectors **a**, **b**, **c**, **d** start from the origin, otherwise be arbitrary. If **a**, **b** and **c** are not coplanar then the endpoints of the quadruple **a**, **b**, **c**, **d** are coplanar if and only if there exist real numbers α, β, γ such that

(1) $$\mathbf{d} = \alpha\mathbf{a} + \beta\mathbf{b} + \gamma\mathbf{c}, \qquad \alpha + \beta + \gamma = 1.$$

Indeed, if the endpoints of the vectors are coplanar then the vectors $\mathbf{a} - \mathbf{d}$, $\mathbf{b} - \mathbf{d}$, $\mathbf{c} - \mathbf{d}$ are lying in the same plane and, accordingly, there exist real numbers λ, γ, ν not all zero such that

$$\lambda(\mathbf{a} - \mathbf{d}) + \gamma(\mathbf{b} - \mathbf{d}) + \nu(\mathbf{c} - \mathbf{d}) = 0.$$

Hence

$$(\lambda + \mu + \nu)\mathbf{d} = \lambda\mathbf{a} + \mu\mathbf{b} + \nu\mathbf{c}.$$

Since **a**, **b**, **c** are not coplanar, $\lambda + \mu + \nu \neq 0$, and thus with

$$\alpha = \frac{\lambda}{\lambda + \mu + \nu}, \qquad \beta = \frac{\mu}{\lambda + \mu + \nu}, \qquad \gamma = \frac{\nu}{\lambda_\mu + \nu}$$

we get the desired result. Since the steps of the argument can be reversed the proof is complete.

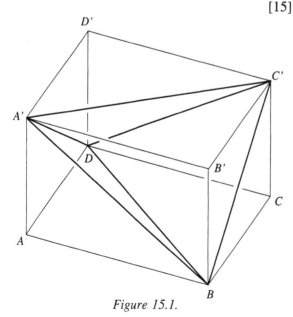

Figure 15.1.

[15] *Circumscribed parallelepiped.* The endpoints of the non parallel diagonals of two opposite (parallel) faces of a parallelepiped fom a tetrahedron. The edges of this tetrahedron are the face diagonals of the prism and opposite edges are lying on oppposite faces. The prism itself is the circumscribed parallelepiped of the tetrahedron. (*Figure 15.1*)

One can, in fact, construct a circumscribed parallelepiped about any tetrahedron by laying a plane through each edge parallel to the opposite edge; the circumscribed parallelepiped of the tetrahedron is bounded by the planes hence obtained.

Certain properties of the tetrahedron become transparent when one takes the circumscribed parallelepiped into account. The centroid of a tetrahedron, for example, is incident to the centre of this parallelepiped. The circumscribed parallelepiped of an equilateral tetrahedron is a cuboid, since as the opposite edges of the tetrahedron, the diagonals are equal on each face.

16] *Hero's formula.* Hero's formula expresses the area of a triangle in terms of its sides. There are several ways to draft it:

$$A^2 = s(s-a)(s-b)(s-c) = \frac{1}{16}(a+b+c)(-a+b+c)(a-b+c)(a+b-c) =$$

$$= \frac{1}{16}\left[(a+b)^2 - c^2\right]\left[c^2 - (a-b)^2\right] = \frac{1}{16}(2a^2b^2 + 2b^2c^2 + 2c^2a^2 - a^4 - b^4 - c^4) =$$

$$= \frac{1}{16}\begin{vmatrix} 0 & 1 & 1 & 1 \\ 1 & 0 & c^2 & b^2 \\ 1 & c^2 & 0 & a^2 \\ 1 & b^2 & a^2 & 0 \end{vmatrix}.$$

17] *Specific points of an equilateral tetrhedron*   see [11].

18] *Convex hull.* The convex hull of a set S of points is the convex set containing S and contained by each convex set that contains S. To put it simply the convex hull is the "smallest" convex set containing S,

There is a nice way to visualize the convex hull of a finite planar set: imagine that the points are nails driven in a table and stretch an elastic ribbon about them: the convex hull is the region bounded by the elastic.

.   Convex hulls in the plane (in the space) can be obtained as the intersections of halfplanes (halfspaces) containing the set S.

19] *Tangent segments in a triangle.* A frequently used fact in elementary arguments that the points of contacts of the circles that are touching the sides of a triangle are dividing them into segments whose lengths can be expressed in a simple manner in terms of the sides of the triangle. The underlying elementary fact is that the tangents to a circle from an external point are congruent. The review of their respective lengths can be checked on the *Figures 19.1* and *19.2*.

20] *Orthocentric tetrahedron.* The lines from the vertices of a tetrahedron perpendicular to the opposite faces are the altitudes of the tetrahedron. If they meet at a common point then this point is called the orthocentre of the tetrahedron and the tetrahedron itself is orthocentric.

Here there are but a few properties of orthocentric tetrahedra:

a) the opposite edges are perpendicular;

b) the sum of the squares is the same for each pair of opposite edges;

c) the faces of their circumscribed parallelepiped are rhombs;

d) their centroid, orthocentre and the centre of the circumscribed sphere are collinear (Euler line).

Conditions a), b), c) are also sufficient for a tetrahedron to be orthocentric; moreover, it is enough to assume that the condition in either a) or b) holds for two opposite pairs of edges only.

[21] *Euler's totient function; Euler's congruence theorem.* The number of the non negative integers up to m that are coprime to m is is the totient function of Euler; its value is denoted by $\varphi(m)$. Here there are a few of its important properties:

1. $a^{\varphi(m)} \equiv 1 \,(\mathrm{mod}\ m)$ if a is a positive integer and $(a, m) = 1$. This is Euler's congruence theorem.

2. If $(a, b) = 1$ then $\varphi(ab) = \varphi(a) \cdot \varphi(b)$.

3. If m is written as the product primes: $m = p_1^{\alpha_1} p_2^{\alpha_2} \ldots p_r^{\alpha_r}$ then,

$$\varphi(m) = m \left(1 - \frac{1}{p_1}\right) \left(1 - \frac{1}{p_2}\right) \ldots \left(1 - \frac{1}{p_r}\right) \quad \text{and } \varphi(1) = 1.$$

[22] *Cauchy's inequality.* Let $(a_1, a_2, \ldots, a_n)$ and $(b_1, b_2 \ldots, b_n)$ be n-tuples of real numbers ("n-dimensional vectors"); then

$$(a_1 b_1 + a_2 b_2 + \ldots + a_n b_n)^2 \le \left(a_1^2 + a_2^2 + \ldots + a_n^2\right) \left(b_1^2 + b_2^2 + \ldots + b_n^2\right).$$

This inequality is often stated as

$$a_1 b_1 + a_2 b_2 + \ldots + a_n b_n \le \sqrt{a_1^2 + a_2^2 + \ldots + a_n^2} \sqrt{b_1^2 + b_2^2 + \ldots + b_n^2}.$$

If the numbers b_i are not all zero then equality holds if and only if there exists a real number $\lambda \ne 0$ such that $a_i = \lambda b_i$ $(i = 1, 2, \ldots, n)$ (The n-dimensional vectors are "parallel").

Its simplest proof is as follows: if the numbers b_i are not all zero then the quadratic

$$(a_1 - \lambda b_1)^2 + (a_2 - \lambda b_2)^2 + \ldots + (a_n + \lambda b_n)^2 = 0$$

in λ has a solution if $a_i = \lambda b_i$ (for every i). Since there can be no more than one solution, its discriminant is not positive and this is exactly the claim. If, on the other hand, the numbers b_i are all zero, then the inequality is obvious.

[23] *Groups.* One of the most frequently occuring algebraic structures. A set of elements forms a group if there is a law of composition which when acting on arbitrary two elements on a definite order assigns an element of the set to this pair. This operation is usually referred to as group multiplication and it is denoted as the ordinary multiplication of numbers. If, for example, a and b are two elements of the group then ab is their product. This operation has the following properties:

1. If a, b, c are the elements of the group then $(ab)c = a(bc)$ (associativiy);

2. There exists an element e of the group such that for any element a

$$ea = ae = a.$$

(e is the neutral element of the group);

3. For any element a of the group there exists an element denoted by a^{-1} such that

$$aa^{-1} = a^{-1}a = e.$$

a^{-1} is the inverse of a.

If $ab = ba$ holds for any two elements, the operation is commutative, then the group is called abelian and the binary operation is then called – and denoted – addition.

Examples

1.The set of integers under addition. Zero is the neutral element and the inverse of each integer is its opposite.

2. The set of real numbers when zero is excluded under multiplication.The neutral element is 1 and the inverse of every element is its reciprocal.

3. The rotations mapping a regular hexagon into itself. The operation is the composition of rotations. The neutral element is the identity transformation and the inverse of the rotation by α is the rotation by $-\alpha$. This is a finite non abelian group.

[24] *Ptolemy's theorem.* In its general form it states that for the opposite sides a, c and b, d and the diagonals e, f of a convex quadrilateral

$$ac + bd \geq ef,$$

and equality holds if and only if the quadrilateral is cyclic. In the proof we shall refer to the notations of *Figure 24.1.*

Apply, for the triangle DAB a rotation and enlargement about D which maps it into the triangle DCB'. The scale factor of the enlargement is $\frac{c}{d}$ and thus $DB' = \frac{ec}{d}$ and $CB' = \frac{ac}{d}$. The triangles ADC and BDB' are similar because they are mathcing in their angle at D and also in the ratio of the neighbouring sides; the scale factor of this similarity is $\frac{e}{d}$ and hence $BB' = \frac{f \cdot e}{d}$.

The triangle inequality for the triangle BCB' yields

$$b + \frac{ac}{d} \geq \frac{ef}{d}, \qquad \text{that is} \qquad ac + bd \geq ef.$$

Equality holds if and only if C is lying on the segment BB' that is $\alpha + \gamma = 180°$, the quadrilateral is indeed cyclic.

[25] *Equilateral cones.* A circular cone is called equilateral if it has three pairwise perpendicular generators.

We are going to prove the following theorem:

If a circular cone has three pairwise perpendicular generators then there are infinitely many such triples, moreover, every generator is the member of such a perpendicular triple.

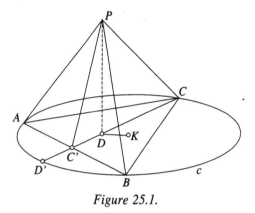

Figure 25.1.

Denote the apex of the cone by P and its base circle by c.

Let the three pairwise perpendicular generators be PA, PB, PC and denote the perpendicular projection of P on the plane of c by D (*Figure 25.1.*). It is easy to prove (Pithagoras, for example) that the triangle ABC is acute angled. Next we show that D is the orthocentre of the triangle ABC. This would follow if the line connecting D to any vertex of the triangle is perpendicular to the opposite side. Consider the vertex C, for example. The side AB is perpendicular to PC because the latter is perpendicular to the plane ABP (it is, in fact, perpendicular to two straight lines lying in this plane) and thus AB is perpendicular to every straight line in the plane ABP. On the other hand, AB is perpendicular to PD since the latter is perpendicular to the plane of c. Therefore, AB is perpendicular to two intersecting lines both lying in the plane PDC and thus it makes a right angle with any line in this plane, CD in particular, therefore, CD is an altitude, indeed.

Denote the intersection of CD with AB by C' and its second intersection with the circle c by D'. Since the mirror image of the orthocentre in a side is incident to the circumcircle, $DC' = C'D'$. In the right triangle $C'PC$ the altitude to the hyptenuse is $PD = h$ and thus, by the geometric mean theorem

(1) $$PD^2 = h^2 = C'D \cdot DC = \frac{1}{2}D'D \cdot DC.$$

The product $DD' \cdot DC$ is the power of the point D with respect to the circle c. (See [6].) This product is hence independent of C's choice.

Choose now an arbitrary point C_1 on the circle and let C_1D meet the circle at D_1. The perpendicular bisector of DD_1 meets the circle at A_1 and B_1 and the midpoint of DD_1 is C_1'. We show that the generators PA_1, PB_1 and PC_1 are pairwise perpendicular. Since

$$C'D \cdot DC = C_1D \cdot DC_1,$$
$$PD^2 = C_1'D \cdot DC_1,$$

by (1), the triangle $C_1'PC_1$ is right angled (by the converse of the geometric mean theorem). Since A_1B_1 is perpendicular to both C_1D and PD, it is also perpendicular to their plane and hence to PC_1 and thus, by the same way, since PC_1 is perpendicular to PC_1', it is also perpendicular to PA_1 and PB_1.

We are left to prove that PA_1 and PB_1 are also perpendicular. Note first, for the proof, that D has to be the orthocentre of the triangle $A_1B_1C_1$. Indeed, the orthocentre is the only one point on the altitude from C_1 whose mirror image in the side is lying on the circumcircle and by the construction of the points A_1B_1 the point D now has this property. Similarly, we obtain that PA_1 is perpendicular to PC_1 and also to PB_1. We have thus shown that any generator of the cone belongs to some family of pairwise perpendicular triple of generators.

Circular cones of this property are called equilateral. It follows from the proof that for a given circle c and point P whose distance from the plane of the circle is h and whose perpendicular projection on the base plane is D the point P is the apex of an equilateral cone of directrix c if and only if PD^2 is equal to the half of the power of D with respect to c.

In order to prove that in problem No. 2 of 1978 there are in fact infinitely many cuboids whose vertex opposite to P is Q we have to show that for a given point P and a circle c the vectors $\overrightarrow{PA}$, $\overrightarrow{PB}$, $\overrightarrow{PC}$ are spanning a cuboid whose vertex opposite to P is Q. This would follow had we shown that the sum $\overrightarrow{PA} + \overrightarrow{PB} + \overrightarrow{PC}$ is constant.

Let the vector from the centre K of c to D be $\mathbf{d}$ (this vector is clearly constant). Now

$$\overrightarrow{PA} + \overrightarrow{PB} + \overrightarrow{PC} = 3\overrightarrow{PD} + \overrightarrow{DA} + \overrightarrow{DB} + \overrightarrow{DC} =$$
$$= 3\overrightarrow{PD} + \left(\overrightarrow{KA} - \mathbf{d}\right) + \left(\overrightarrow{KB} - \mathbf{d}\right) + \left(\overrightarrow{KC} - \mathbf{d}\right) =$$
$$= 3\overrightarrow{PD} - 3\mathbf{d} + \left(\overrightarrow{KA} + \overrightarrow{KB} + \overrightarrow{KC}\right).$$

It is well known that the sum of the vectors from the circumcentre to the vertices of a triangle is leading to the orthocentre. Therefore, $\overrightarrow{KA} + \overrightarrow{KB} + \overrightarrow{KC} = \mathbf{d}$ and thus

$$\overrightarrow{PA} + \overrightarrow{PB} + \overrightarrow{PC} = 3\overrightarrow{PD} - 2\mathbf{d},$$

and this sum does not depend on the choice of the points A, B, C indeed.

[26] *A representation of positive integers.* In the solution of the problem No. 3. of 1978 we have used the following theorem: if α and β are positive irrational numbers satisfying $\dfrac{1}{\alpha} + \dfrac{1}{\beta} = 1$ then the sequences

$$\{[n\alpha]\}, \qquad [\{n\beta\}] \qquad n = 1, 2, \ldots$$

have no common elements and together they exhibit every positive integer.

Note first that the two numbers α and β are greater than 1 and thus the sequences $[n\alpha]$, $[n\beta]$ are strictly increasing. We now prove that, for any positive integer N there is either a positive integer k such that $[k\alpha] = N$, or a positive integer m such that $[n\beta] = N$, moreover, the two options cannot hold at the same time. There clearly exist the unique positive integers k and m such that

$$[(k-1)\alpha] < N \le [k\alpha], \qquad \text{and}$$
$$[(m-1)\beta)] < N \le [m\beta].$$

Therefore,

$$k\alpha - \alpha < N < k\alpha,$$
$$m\beta - \beta < N < m\beta.$$

Subtracting N from each term of the above inequalities:

$$(k\alpha - N) - \alpha < 0 < k\alpha - N,$$
$$(m\beta - N) < 0 < m\beta - N.$$

Introducing the notations $d = k\alpha - N$ and $d' = m\beta - N$ we get

(1) $\qquad 0 < d < \alpha, \quad 0 < d' < \beta, \quad \text{that is} \quad 0 < \dfrac{d}{\alpha} < 1, \quad 0 < \dfrac{d'}{\beta} < 1.$

Since $k = \dfrac{N}{\alpha} + \dfrac{d}{\alpha}$, $m = \dfrac{N}{\beta} + \dfrac{d'}{\beta}$, (1) implies

$$k + m = N\left(\frac{1}{\alpha} + \frac{1}{\beta}\right) + \frac{d}{\alpha} + \frac{d'}{\beta} = N + \frac{d}{\alpha} + \frac{d'}{\beta},$$

that is

(2) $\qquad \dfrac{d}{\alpha} + \dfrac{d'}{\beta} = k + m - N.$

By (1), on the other hand

$$0 < k + m - N < 2.$$

Since k, m, N are positive integers, this implies

$$k + m - N = 1.$$

Now by (2)

$$\frac{d}{\alpha} + \frac{d'}{\beta} = 1 = \frac{1}{\alpha} + \frac{1}{\beta},$$

that is $\qquad\qquad\qquad \alpha(d' - 1) = (1 - d)\beta.$

Since α and β are positive and d is irrational, this equality implies that one of d and d' is less than 1 and the other one is greater than 1. If, for example, $d < 1$, $d' > 1$ then, by the definition of d and d' implies

$$\alpha k = N + d, \qquad\qquad [\alpha k] = N,$$
$$\beta m = N + d' \qquad\qquad [\beta m] > N,$$

and thus N belongs to exactly one of the sequences $[\alpha k]$ and $[\beta m]$. The same holds if $d > 1$ and $d' < 1$.

7] *Solving linear recurrences.* To find a formula for the nth term of a sequence defined by recurrence relations is always an important task. Here we present the solution of the following special case of the problem.

$$(1) \qquad a_n = c_1 a_{n-1} + c_2 a_{n-2}$$

This is a so called second order linear recurrence, the coefficients a_1 and a_2 are given numbers.

The heart of the matter is to find geometric progressions satisfying (1) and unfold the general solution as a linear combination of these particular sequences. Here there is the method. The quadratic

$$x^2 - c_1 x - c_2 = 0$$

is called the *characteristic equation* to the recurrence (1). Denote its roots (real or complex) by x_1 and x_2.

Assume, first, that $x_1 \neq x_2$. Then the nth term of the sequence is equal to

$$(2) \qquad a_n = \lambda x_1^n + \mu x_2^n,$$

where λ and μ are constants depending on the initial terms of the sequence; their actual value can be computed by solving the simultaneous system

$$(3) \qquad \begin{aligned} \lambda x_1 + \mu x_2 &= a_1, \\ \lambda x_1^2 + \mu x_2^2 &= a_2. \end{aligned}$$

If $x_1 = x_2$ then the nth term can be computed as

$$a_n = \lambda x_0^2 + \mu n x_0^{n-1}$$

(we have adopted the notation $x_1 = x_2 = x_0$). The system for the values of λ and μ is now

$$\lambda x_0 + \mu = a_1,$$

$$\lambda x_0^2 + 2\mu x_0 = a_2.$$

The method works essentially the same way in the general case. Consider the sequence $\{a_i\}$ defined by

$$(4) \qquad a_{n+k} = c_1 a_{n+k+1} + c_2 a_{n+k+2} + \ldots + c_k a_n.$$

This relation is called k-order linear recurrence with constant coefficients. The numbers c_i are constants and there are also given the initial values $a_1, a_2, \ldots, a_k$. The characteristic polynomial corresponding to the above recurrence is

$$x^k - c_1 x^{k-1} - c_2 x^{k-2} - \ldots - c_k = 0.$$

If its roots (real or complex) are $x_1, x_2, \ldots, x_k$ are distinct then the terms of the sequence $\{a_i\}$ can be computed as

$$(5) \qquad a_n = \lambda_1 x_1^{n-1} + \lambda_2 x_2^{n-1} + \ldots + \lambda_n x_k^{n-1}$$

where the coefficients $\lambda_1, \lambda_2, \ldots, \lambda_k$ can be calculated from the initial values of the sequence.

The method still works if there happen to be multiple roots of the characteristic polynomial, although, as in the second order case, the actual solution is a bit more tedious.

[28] *Two relations among binomial coefficients.*

A) $\binom{n}{k} = \binom{n-1}{k} + \binom{n-1}{k-1}$;

B) $\binom{n+1}{k+1} = \binom{n}{k} + \binom{n-1}{k} + \ldots + \binom{k}{k}$.

The proof of A) is straightforward from the defining equality $\binom{n}{k} =$

$$= \frac{n!}{k!(n-k)!}.$$

$$\binom{n-1}{k} + \binom{n-1}{k-1} = \frac{(n-1)!}{k!(n-k-1)!} + \frac{(n-1)!}{(k-1)!(n-k)!} =$$

$$= \frac{(n-1)!(n-k+k)}{k!(n-k)!} = \binom{n}{k}.$$

To prove B) one can use A):

$$\binom{n+1}{k+1} = \binom{n}{k+1} + \binom{n}{k},$$

$$\binom{n}{k+1} = \binom{n-1}{k+1} + \binom{n-1}{k},$$

$$\binom{n-1}{k+1} = \binom{n-2}{k+1} + \binom{n-2}{k},$$

$$\vdots$$

$$\binom{n-(n-k-2)}{k+1} = \binom{k+2}{k+1} = \binom{k+1}{k+1} + \binom{k+1}{k}.$$

Summing the equalities and considering that $\binom{k+1}{k+1} = \binom{k}{k}$ one arrives to the claim.

[29] *Menelaus' theorem.* Let C_1, A_1 and B_1 be points on the sides AB, BC, CA of the triangle ABC, respectively. These points are collinear if and only if

(1) $$\frac{AC_1}{C_1B} \cdot \frac{BA_1}{A_1C} \cdot \frac{CB_1}{B_1A} = -1.$$

As for the sign of the fractions on the l. h. s. they are positive if the vectors $\overrightarrow{AC_1}$ and $\overrightarrow{C_1B}$ are oriented similarly, otherwise they are negative.

Note here that disregarding the orientation of the segments there is 1 on the r. h. s. of (1) and this form of the claim is just necessary for the points A_1, B_1, C_1 to be collinear.

[30] *Residue classes, congruences.* For a given integer $m > 1$ the integers a and b are said to belong to the same residue class "with respect to m", or simply *modulo* m ("mod m", for short) if they give equal remainders when divided by m. Since the possible remainders are 0, 1, 2, ..., $m - 1$, there are m residue classes with respect to m and every integer belongs to exactly one of them, or, putting it differently, every whole number is representing some residue class.

m integers form a so called *complete residue system* mod m if they represent distinct residue classes; together they hence represent every possible remainder mod m.

Two integers clearly belong to the same residue class mod m if their difference is divisible by m. For given integers a and b this is denoted as

$$a \equiv b \quad (\text{mod } p) \qquad \text{or simply} \qquad a \equiv b \quad (m).$$

This is the relation of congruence. Several properties of this relation are resembling to those of equality; here there are a few of them. (For sake of brevity the mod m extension is now omitted.)

1. if $a \equiv b$, akkor $b \equiv a$; $a \equiv a$ for every integer a;

2. if $a \equiv b$ and $b \equiv c$ then $a \equiv c$;

3. if $a \equiv b$ and $c \equiv d$ then

$$a + c \equiv b + d, \quad a - c \equiv b - d, \quad ac \equiv bd, \quad a^n \equiv b^n, \quad (n \text{ is a positive integer});$$

4. if $ac \equiv bc$ then $a \equiv b \left(\text{mod } \dfrac{m}{d} \right)$, where $d = (c, m)$).

[31] *Designs.* The v element set H is called block design if there exist b subsets in H of k elements each – the blocks – every element of H belongs to exactly r blocks and any two distinct elements of H belong to exactly λ blocks.

The numbers v, b, k, r, λ are the *parameters* of the block design and they are related in several ways. (Obviously $2 \leq k < v$ and $\lambda > 0$). Assign, to the elements of the block design the rows of a matrix of v rows and b columns and its columns to the blocks as follows: a given entry of the matrix is equal to 1 if the element corresponding to its row belongs to the block corresponding to its column; otherwise the entry is equal to zero. The matrix hence obtained is the so called incidence matrix of the block design. There are exactly r copies of 1 in each row and k 1-s in each column. Tallying the 1-s both row and columnwise yields

$$bk = vr.$$

Counting further incidences one gets

$$r(k-1) = \lambda(v-1).$$

To find the proper conditions under which there exists a block design for a given system of parameters is one of the unsolved hard problems of combinatorics. If, for example, $v = b = p^{2\alpha} + p^\alpha + 1$, $\lambda = 1$, $k = r = p^\alpha + 1$, where p is a prime and α is a positive integer then one can construct the corresponding block design, these are the so called finite projective planes.

[32] *Fermat's congruence theorem (the little "Fermat's theorem")*. For any prime p the number of coprimes to p below p is equal to $p-1$, $\varphi(p) = p-1$ and by Euler's congruence theorem [21]

$$a^{p-1} \equiv 1 \pmod{p}, \qquad \text{if } (a,p) = 1.$$

This is Fermat's congruence theorem. Multiplying both sides by a yields

$$a^p \equiv a \pmod{p}.$$

This form of the theorem holds even if a is not prime to p since as a prime, p then divides a.

[33] *Erdős–Mordell inequality*. Denote the distances of a point P of a triangle from the vertices by p, q and r, respectively, and the perpendicular distances of P from the sides by x, y and z, respectively. Then one has the following inequality

$$p + q + r \geq 2(x + y + z)$$

and equality holds if and only if the triangle is equilateral and P is its centre. This is the Erdős–Mordell-inequality and we are going to prove it.

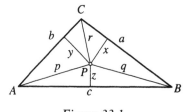

Figure 33.1.

Assume that p, q, r are the distances from the vertices A, B, C and, x, y, z are those from the sides BC, CA, AB, respectively (*Figure 33.1.*).

First we prove the following inequalities:

(1) $ap \geq bz + cy$; $\quad bq \geq cx + az$; $\quad cr \geq zy + bx$.

It is clearly enough to show the first one. Reflect, for the proof, the vertices B and C in the bisector of the angle A and denote the mirror images by C' and B', respectively; the point P remains fixed. The sides of the triangle $AB'C'$ are now $AB' = c$, $B'C' = a$, $C'A = b$; the distances of P from the sides $B'C'$, $C'A$, AB' are x', y, z, respectively (*Figure 33.2.*). The quantity x' refers to signed distance: it is zero, if P is incident to $B'C'$ and if P is outside the triangle $A'B'C'$ then x' is negative.

Denote the altitude from A of the triangle $AB'C'$ by h_a; this is clearly not longer than the path $p+x'$ from A to $B'C'$ through P. Hence

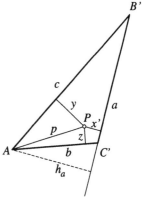

(2) $$m_a \leq p+x'.$$

Multiplying both sides by a and considering that both ah_a and $ax' + bz + cy$ are equal to the double area of the triangle $AB'C'$

$$am_a \leq ap + ax';$$
$$ax' + bz + cy \leq ap + ax',$$

Figure 33.2.

and this implies the claim $ap \geq bz + cy$; a similar argument yields the other two inequalities in (1). Note that the relations hold even if $x' \leq 0$.

Rearranging the inequalities just proved:

$$p \geq \frac{b}{a}z + \frac{c}{a}y;$$
$$q \geq \frac{c}{b}x + \frac{a}{b}z;$$
$$r \geq \frac{a}{c}y + \frac{b}{c}x.$$

The sum of these inequalities gives

(3) $$p+q+r \geq \left(\frac{c}{b}+\frac{b}{c}\right)x + \left(\frac{c}{a}+\frac{a}{c}\right)y + \left(\frac{b}{a}+\frac{a}{b}\right)z.$$

Since the sum of a positive number and its reciprocal is at least 2, one arrives to

$$p+q+r \geq 2(x+y+z).$$

As for equality it must also hold in both (2) and (3). In the latter it holds if and only if $a = b = c$; the sum of a positive number and its reciprocal is 2 only if this number is equal to 1. Our triangle hence must be equilateral.

If this is the case, however, then the triangles ABC and $AB'C'$ in the first paragraph are identical. Hence the position of the segments AP and x' must be identical to that of AP and x; finally, h_a is equal to $(p+x)$ if P is incident to the altitude from A.

This, of course, must hold for the other two altitudes, as well and thus, in case of equality, P must be the intersection of the altitudes, the centre of the equilateral triangle.

[34] *Brocard points.* The points Q_1 and Q_2 are called the Brocard points of the triangle ABC if

$$BAQ_1\angle = CBA_1\angle = ACQ_1\angle, \quad \text{or} \quad ABQ_2\angle = BCQ_2\angle = CAQ_2\angle.$$

These points can be constructed; Q_1, for example, is the intersection of two circles: one of them is passing through A and B and it is touching the line BC and the other one is passing through B and C and touching the line AC. The angles at Q_1 and Q_2 are also equal to each other and

$$\cot \omega = \cot \alpha + \cot \beta + \cot \gamma$$

for their common measure ω. (This is straightforward from the sine rule, for example.)

Apart from he circumcentre of a triangle H the Brocard points are those that have the following property: the feet of the perpendiculars to the sides from these points form a triangle similar to H. This implies that no matter how we rotate the triangle about any one of its Brocard points, the pairwise intersections of the corresponding sides of the two triangles – H and the rotated one – form a triangle that is also similar to H.

[35] *A common origin of certain inequalities.* There is a fundamental inequality showing up in the solutions from time to time. It has various formulations and there are several further inequalities that can be deduced from it. It states that

let $a_1, a_2, \ldots, a_n$ and $b_1, b_2, \ldots, b_n$ are real numbers and $b_{i_1}, b_{i_2}, \ldots, b_{i_n}$ is an arbitrary rearrangement of the numbers b_i. Prepare the sum

$$S = a_1 b_{i_1} + a_2 b_{i_2} + \ldots + a_k b_{i_k}.$$

This sum is maximal if and only if the ordering of the numbers a_i and b_{i_k} is the same and it is minimal if the two orderings are opposite.

Denote, for he proof, the highest terms of the two n-tuples by a_r and b_s, respectively and consider the sum

$$Q = a_1 b_1 + \ldots + a_r b_r + \ldots + a_s b_s + \ldots + a_n b_n.$$

Swap now the two numbers b_r and b_s; if $r \neq s$ then

$$Q' = a_1 b_1 + \ldots + a_r b_s + \ldots + a_s b_r + \ldots + a_n b_n.$$

$$Q - Q' = a_r b_s + a_s b_r - a_r b_r - a_s b_s = (a_r - a_s)(b_s - b_r) \geq 0.$$

$Q' = Q$ holds if and only if $a_r = a_s$ or $b_r = b_s$, but then the highest a_i is, in fact, multiplied by the highest b_i. Through an appropriate sequence of swaps one arrives to similarly ordered n-tuples. Since the sum Q is not decreasing, the maximum is attained when the orderings are the same, indeed. The corresponding statement about the minimum follows similarly.

[36] *Four circles theorem.* Consider four straight lines whose pairwise intersections are distinct. The circumcircles of the four triangles hence obtained are passing through a common point S (*Figure 36.1.*)

Denote, as in the diagram, the intersection of the circumcircles of the triangles ABC and CDE by S. It is clearly enough to show, by symmerty, that the circumcircle of the triangle ADF is, in fact, passing through S. Intercepted by the same arcs $SDE\angle = SCE\angle$ and in the cyclic quadrilateral $ABCS$ this angle is equal to $SAF\angle$, the quadrilateral $ASDF$ is cyclic and thus the circumcircle of the triangle ADF is passing through S, indeed.

By the Simson theorem the feet of the perpendiculars from S to the four lines are collinear. Of what we know about conic sections, it follows that four lines as tangents to the curve determine a unique parabola; the feet of the perpendiculars from the focus to the tangents are on the tangent through the vertex; the circumcircles of the triangles formed by three tangents to the parabola are passing through the focus. Taking these facts into account we get that the point S is, in fact, the focus of the parabola determined by the four straight lines.

[37] *Radius inequality.* The diameter of the incircle of a triangle cannot exceed the circumradius, that is

$$R \geq 2r.$$

This is an immediate consequence of Euler's identity $d^2 = R^2 - 2Rr$ (d is the distance of the two centres), but there are several proofs around. It is also related to various triangle inequalities and relations, for example

$$\cos \alpha + \cos \beta + \cos \gamma \leq \frac{3}{2},$$

$$\cos \alpha + \cos \beta + \cos \gamma = 1 + \frac{r}{R},$$

$$(-a+b+c)(a-b+c)(a+b-c) \leq abc.$$

The equality $R = 2r$ holds only in equilateral triangles.

[38] *Parallel chords.* Consider a circle about the origin on the Argand diagram and denote its four points by the complex numbers a, b, c and d. The chords connecting a to b and c to d are parallel if

$$ab = cd.$$

Assume that the counterclockwise order of the points is a, b, c and d. The corresponding chords are then parallel if and only if the arcs $\overgroup{b,c}$ and $\overgroup{d,a}$ are equal. Denote the central angle of these arcs by φ and let $e = \cos \varphi + i \sin \varphi$. Multiplication by e is hence a rotation by φ about the origin.

$$be = c \quad \text{and} \quad de = a,$$

from which

.

$$\frac{be}{de} = ca, \qquad \frac{b}{d} = \frac{c}{a}, \qquad \text{that is} \qquad ab = cd,$$

and the converse of the argument is also valid.

[39] *n-dimensional vectors.* Ordered n-tuples $(a_1, a_2, \ldots, a_n)$ of real numbers are sometimes called n-dimensional vectors and they are then denoted a single bold face letter: $\mathbf{a}(a_1, a_2, \ldots, a_n)$; the numbers a_i are then the coordinates of the vector. There is a natural way to perform algebraic operations between n-dimensional vectors as follows:

addition: the sum of the vectors $\mathbf{a}(a_1, a_2, \ldots, a_n)$ and $\mathbf{b}(b_1, b_2, \ldots, b_n)$ is

$$\mathbf{a} + \mathbf{b}(a_1 + b_1, a_2 + b_2, \ldots, a_n + b_n).$$

Subtraction: $\mathbf{a} - \mathbf{b}(a_1 - b_1, a_2 - b_2, \ldots, a_n - b_n).$

Multiplication by the real number λ: $\lambda \mathbf{a}(\lambda a_1, \lambda a_2, \ldots, \lambda a_n).$

Scalar, or dot product: $\mathbf{ab} = a_1 b_1 + a_2 b_2 + \ldots + a_n b_n.$

The fundamental algebraic laws are as follows:

$$\mathbf{a} + \mathbf{b} = \mathbf{b} + \mathbf{a}, \qquad \mathbf{ab} = \mathbf{ba}, \qquad \lambda \mathbf{a} = \mathbf{a}\lambda, \qquad \lambda(\mu \mathbf{a}) = (\lambda\mu)\mathbf{a} = \lambda\mu\mathbf{a},$$

$$(\lambda + \mu)\mathbf{a} = \lambda\mathbf{a} + \mu\mathbf{a}, \qquad \lambda(\mathbf{a} + \mathbf{b}) = \lambda\mathbf{a} + \lambda\mathbf{b}, \qquad \lambda(\mathbf{ab}) = (\lambda\mathbf{a})\mathbf{b} = \mathbf{a}(\lambda\mathbf{b}),$$

$$\mathbf{a}(\mathbf{b} + \mathbf{c}) = (\mathbf{b} + \mathbf{c})\mathbf{a} = \mathbf{ab} + \mathbf{ac} \qquad \text{(distributive law)}.$$

Further notations and concepts: $\dfrac{\mathbf{a}}{\lambda} = \dfrac{1}{\lambda}\mathbf{a}$. The vector $\mathbf{0}(0, 0, \ldots, 0)$ is called zero vector; the product of two equal vectors is called the square of the given vector and abbreviated accordingly: $\mathbf{aa} = \mathbf{a}^2 = a_1^2 + a_2^2 + \ldots + a_n^2.$

The distributive law also holds if both factors have several terms; in particular:

$$(\mathbf{a}_1 + \mathbf{a}_2 + \ldots + \mathbf{a}_n)^2 = a_1^2 + a_2^2 + \ldots + a_n^2 + 2(\mathbf{a}_1\mathbf{a}_2 + \mathbf{a}_1\mathbf{a}_3 + \ldots + \mathbf{a}_{n-1}\mathbf{a}_n).$$

The proof of any one of the above relations can be done by expanding them in terms of coordinates. For a further application see also [22].

[40] *Weighed means.* The notion of weighed means is a generalization of the notion of means.

Assign, as its weight, to each of the real numbers $a_1, a_2, \ldots, a_n$ a positive number s_i, respectively. The weighed arithmetic mean (or weighed average) of the numbers $a_1, a_2, \ldots, a_n$ is then

$$A_s = \frac{s_1 a_1 + s_2 a_2 + \ldots + s_n a_n}{s_1 + s_2 + \ldots + s_n};$$

their weighed geometric mean is

$$G_s = {}^{s_1 + s_2 + \ldots + s_n}\!\!\sqrt{a_1^{s_1} a_2^{s_2} \ldots a_n^{s_n}};$$

the weighed harmonic mean is

$$H_s = \frac{s_1 + s_2 + \ldots + s_n}{\frac{s_1}{a_1} + \frac{s_2}{a_2} + \ldots + \frac{s_n}{a_n}},$$

and, finally, the weighed quadratic mean is

$$Q_s = i \frac{s_1 a_1^2 + s_2 a_2^2 + \ldots + s_n a_n^2}{s_1 + s_2 + \ldots + s_n}.$$

Togeteher these weighed means also obey the well known chain of inequalities between ordinary means, namely

$$H_s \leq G_s \leq A_s \leq Q_s.$$

The proof follows a general pattern. The first step it is straightforward: if the weights s_i are whole numbers then there is nothing new here, the weighed means can be conceived as ordinary means of appropriate number of copies of each number: there are s_i occurences of the number a_i. If the weights are rational then, as the following example shows, the issue can be reduced to the previous case. Let's see how to do this in the $A_s \geq G_s$ inequality for two terms; let the weights be $s_1 = \frac{p_1}{q_1}$ and $s_2 = \frac{p_2}{q_2}$ (p_i and q_i are positive integers). Then

$$A_s = \frac{\frac{p_1}{q_1} a_1 + \frac{p_2}{q_2} a_2}{\frac{p_1}{q_1} + \frac{p_2}{q_2}} = \frac{p_1 q_2 a_1 + p_2 q_1 a_2}{p_1 q_2 + p_2 q_1} \geq \sqrt[p_1 q_2 + p_2 q_1]{a_1^{p_1 q_2} \cdot a_2^{p_2 q_1}} =$$

$$= \sqrt[\frac{p_1}{q_1} + \frac{p_2}{q_2}]{a_1^{\frac{p_1}{q_1}} \cdot a_2^{\frac{p_2}{q_2}}} = G_s.$$

Finally, if there happen to be irrational numbers among the weights, then one should invoke standard continuity arguments. The point is that the means are continuous functions of the weights and irrational numbers can be approximated to arbitrary precision by rationals.

[41] *Trigonometric form of Ceva's theorem and an application.* If the lines a', b', c' are dividing the angles α, β, γ of the triangle ABC into the parts α_1 and α_2, β_1 and β_2, γ_1 and γ_2, respectively (*Figure 41.1*) then the lines a', b', c' are concurrent if and only if

(1)
$$\frac{\sin \alpha_1 \sin \beta_1 \sin \gamma_1}{\sin \alpha_2 \sin \beta_2 \sin \gamma_2} = 1.$$

Assume first that the three lines in question are passing through a common point P. By the sine rule in the triangles ABP, BCP, CAP respectively

$$\frac{PA}{PB} = \frac{\sin \beta_1}{\sin \alpha_2}, \qquad \frac{PB}{PC} = \frac{\sin \gamma_1}{\sin \beta_2},$$

$$\frac{PC}{PA} = \frac{\sin \alpha_1}{\sin \gamma_2}.$$

and the product of the three equalities yields (1).

Assume, for the converse, that the lines a', b', c' divide the angles of the triangle according to (1). Denote the intersection of the lines a' and b' by P' and suppose that $P'C$ cuts the angle γ into the parts γ'_1 and γ'_2. We have already seen that

$$\frac{\sin \alpha_1 \sin \beta_1 \sin \gamma'_1}{\sin \alpha_2 \sin \beta_2 \sin \gamma'_2} = 1,$$

which, when compared to (1), yields $\dfrac{\sin \gamma_1}{\sin \gamma_2} = \dfrac{\sin \gamma'_1}{\sin \gamma'_2}$, that is

$$\frac{\sin(\gamma - \gamma_2)}{\sin \gamma_2} = \frac{\sin(\gamma - \gamma'_2)}{\sin \gamma'_2}, \qquad \text{or}$$

$$\sin \gamma'_2 (\sin \gamma \cos \gamma_2 - \cos \gamma \sin \gamma_2) = \sin \gamma_2 (\sin \gamma \cos \gamma'_2 - \cos \gamma \sin \gamma'_2),$$

$$\sin \gamma'_2 \cos \gamma_2 = \sin \gamma_2 \cos \gamma'_2,$$

$$\sin(\gamma'_2 - \gamma_2) = 0.$$

This implies $\gamma_2 = \gamma'_2$ and $\gamma_1 = \gamma'_1$ and thus P' and P are identical, the proof is complete.

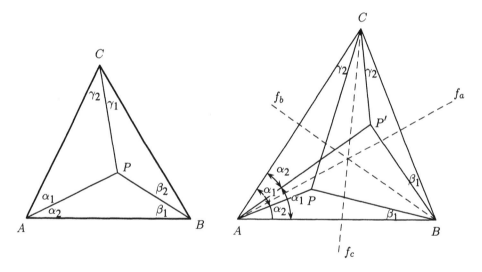

Figure 41.1.   Figure 41.2.

An immediate consequence is the theorem used in the solution of Qu. No 2. in 1996: if P is an interior point of the triangle ABC and the line PA is reflected in the bisector of $A\angle$, PB is reflected to the bisector of $B\angle$, finally PC is reflected in the bisector of $C\angle$, then the reflected lines are concurrent. Indeed, the parts α_1 and α_2, β_1 and β_2, γ_1 and γ_2 are swapped under the reflections and hence (1) remains valid, the mirror lines are also passing through a common point (*Figure 41.2.*).

This assertion can be proved in a more general form without the Ceva-trigonometry machinery using reflections only; if the lines through the vertices of a triangle belong to a pencil (i.e. they are concurrent or parallel) then the same holds for the mirror images in the corresponding angle bisectors.

[42] *An extension of the Erdős–Mordell inequality.* Let P, Q and S be interior points on the sides AB, BC, CA of the triangle ABC, respectively. Denote the intersection of the perpendiculars to AB at P and to BC at Q by Y and similarly, the intersection of the perpendiculars to BC at Q and to CA at S by Z and, finally, the intersection of the perpendiculars to CA at Q and to AB at P by X. If, additionally, the points X, Y and Z are interior to the triangle ABC then

$$AX + BY + CZ \geq XP + YP + YQ + ZQ + ZS + XS,$$

and equality holds if and only if the triangle ABC is equilateral and the each of the points X, Y, Z are at the centre of ABC. (*Figure 42.1.*).

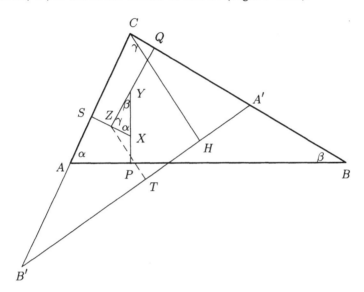

Figure 42.1.

The proof is following the demonstration in [33] of the original theorem. Denote by A', B' the mirror image of the vertices A and B in the interior bisector of the angle C, respectively; the feet of the perpendiculars to $A'B'$ from C and Z be H and T, respectively. The area of the triangle $A'B'C$ can be written in two different ways:

$$2a_{A'B'C} = 2a_{A'B'Z} + 2a_{CA'Z} + 2a_{B'CZ}.$$

With the compulsory notations $AB = A'B' = c$, $BC = B'C = a$, $CA = CA' = b$ this can be put as

(1) $$c \cdot CH = c \cdot ZT + b \cdot ZQ + a \cdot ZS;$$

here the length of ZT is negative if the line $A'B'$ separates the points Z and C. In any case we have the following inequality:

(∗) $$CZ + ZT \geq CH.$$

Hence

$$c \cdot CZ + c \cdot ZT \geq c \cdot CH,$$

$$c \cdot CZ \geq c \cdot CH - c \cdot ZT,$$

which, when combined with (1), implies

$$c \cdot CZ \geq b \cdot ZQ + a \cdot ZS,$$

$$CZ \geq \frac{b}{c} \cdot ZQ + \frac{a}{c} ZS.$$

Similarly

$$AX \geq \frac{c}{a} XS + \frac{b}{a} XP,$$

$$BY \geq \frac{a}{b} YP + \frac{c}{b} YQ.$$

Adding these inequalities

(2) $$AX + BY + CZ \geq \left(\frac{a}{b} YP + \frac{b}{a} XP \right) + \left(\frac{b}{c} ZQ + \frac{c}{b} YQ \right) + \left(\frac{c}{a} XS + \frac{a}{c} ZS \right).$$

Apply now the identity

$$kK + nN = (k+n) \frac{K+N}{2} + (k-n) \frac{K-N}{2}$$

for the expression

$$\frac{a}{b} YP + \frac{b}{a} XP = \left(\frac{a}{b} + \frac{b}{a} \right) \frac{YP+XP}{2} + \left(\frac{a}{b} - \frac{b}{a} \right) \frac{YP-XP}{2}.$$

Observe that their angles being pairwise equal the triangles ABC and XYZ are similar. If λ is the scale factor of similarity then

$$\frac{YZ}{a} = \frac{ZX}{b} = \frac{XY}{c} = \lambda,$$

and also $YP - XP = XY = \lambda c$. By

(**) $$\frac{a}{b} + \frac{b}{a} \geq 2.$$

we now get

$$\frac{a}{b} \cdot YP + \frac{b}{a} \cdot XP \geq YP + XP + \lambda \left(\frac{ca}{2b} - \frac{bc}{2a} \right).$$

Similarly

$$\frac{b}{c} \cdot ZQ + \frac{c}{b} YQ \geq ZQ + YQ + \lambda \left(\frac{ab}{2c} - \frac{ca}{2b} \right),$$

$$\frac{c}{a} \cdot XS + \frac{a}{c} \cdot ZS \geq XS + ZS + \lambda \left(\frac{bc}{2a} - \frac{ab}{2c} \right).$$

By (2) the sum of these inequalities implies the claim.

$$AX + BY + CZ \geq XP + YP + YQ + ZQ + ZS + XS.$$

If there is equality then by (∗∗) $a = b = c$ and by (∗) X, Y and Z are laying on the respective altitudes; summarizing the conditions of equality ABC has to be equilateral and each of X, Y, Z must be at its centre.

43] *A property of equilateral triangles.* Let P be an arbitrary point on the arc AB not containing C of the circumcircle of the equilateral triangle ABC. Then $AP + BP = PC$.

Rotate, about A, the triangle APB by $60°$. If the image is ACP' then P' is on the segment CP because $ACP\angle = ABP\angle$ (inscribed angles intercepting the same arc AP). By the rotation the triangle APP' is equilateral and thus $AP = PP'$; since, on the other hand, $BP = CP'$ we get

$$PC = PP' + CP' = AP + BP,$$

indeed. (*Figure 43.1.*).

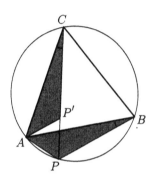

Figure 43.1.

We note that the claim is an immediate consequence of Ptolemy's theorem [24]; when applied to the cyclic quadrilateral $APBC$ it yields

$$AP \cdot BC + BP \cdot CA = AB \cdot PC,$$

and dividing through by the length of the side of the triangle we get the desired result.

44] *The number of divisors.* The number of (positive) divisors of a positive integer n is denoted by $d(n)$. (One can also come across to the notation $\tau(n)$.) If the prime factorization of n is

$$n = p_1^{\alpha_1} \cdot p_2^{\alpha_2} \cdot \ldots \cdot p_r^{\alpha_r},$$

then every divisor of n is equal to

$$p_1^{\beta_1} p_2^{\beta_2} \ldots p_r^{\beta_r}$$

where $0 \leq \beta_i \leq \alpha_i$, and, conversely, for any such choice of the numbers β_i there is a divisor of the above form. Therefore, there are $\alpha_i + 1$ ways to set the value of β_i and, accordingly

$$d(n) = (\alpha_1 + 1)(\alpha_2 + 1) \ldots (\alpha_r + 1)$$

divisors of n, altogether. This shows that $d(n)$ depends on the list of indices only, not on the actual prime factors. As a consequence we note here that if a and b are coprime then

$$d(ab) = d(a)d(b),$$

the function d is multiplicative. This property is, of course, true for products of finitely many coprime factors.

[45] *Turán's graph theorem.* Paul Tur n proved the following theorem in 1941: let $n = q(k-1) + r$, where q, k, r are whole numbers such that $0 \le r < k - 1$. If there are more than

$$E = \frac{k-2}{2(k-1)}(n^2 - r^2) + \binom{r}{2},$$

edges in a simple graph of n vertices then the graph contains a complete subgraph of k vertices. The result is sharp, since for every n there exists a simple graph of n vertices and E edges with no complete subgraph of k vertices.

Milton Keynes UK
Ingram Content Group UK Ltd.
UKHW020703210924
1771UKWH00043B/198

9 781843 312000